Lecture Notes in Mathematics

A collection of informal reports and seminars
Edited by A. Dold, Heidelberg and B. Eckmann, Zürich

T0219890

306

Horst Luckhardt

Philipps-Universität, Marburg an der Lahn/BRD

Extensional Gödel Functional Interpretation

A Consistency Proof of Classical Analysis

Springer-Verlag
Berlin · Heidelberg · New York 1973

AMS Subject Classifications (1970): 02D02, 02E05

ISBN 3-540-06119-3 Springer-Verlag Berlin · Heidelberg · New York
ISBN 0-387-06119-3 Springer-Verlag New York · Heidelberg · Berlin

Offsetdruck: Julius Beltz, Hemsbach/Bergstr.

Preface

The main part of the following Lecture Notes is from the author's Habilitationsschrift presented to Philipps-Universität Marburg/Lahn in November 1970. The final chapter was added in spring 1971; also the introduction was extended giving now a survey of all foundational methods on classical analysis so far developed.

Lectures on the subject were given at the meeting on proof theory at Hannover in October 1971 and at Oberwolfach in spring 1972.

I would like to express my gratitude to professors V.G. Avakumović and K. Schütte as well as to the Deutsche Forschungsgemeinschaft for their support of my proof-theoretic program. Thanks are due to the editors of these Lecture Notes.
Thanks also to Miss Gisela Naumann who typed this English version. The previous German version was typewritten by my wife, Helen Luckhardt-Zimmermann; to her my special thanks.

Marburg/Lahn, September 1972 H. Luckhardt

Contents

0. Introduction and survey

By classical analysis we mean the structure $< N,P(N),0,S,\varepsilon >$ or, more customarily by including functions, $< N,P(N),N^N,0,S,\varepsilon,..(.) >$, where $..(.)$ denotes application between (number-theoretic) functions and (natural) numbers. It is well known that every concrete axiomatic treatment of classical analysis is necessarily incomplete. Therefore we have to concentrate our proof-theoretic study on that part of this structure which is described by the accepted principles of classical mathematics. In the following we understand "(classical) analysis" always in this sense.

The familiar axiomatic second-order characterizations of analysis consist of Peano-axioms, extensionality, comprehension and - in the presence of functions - of recursion equations and axioms of choice in the forms: ω-choice of numbers or of functions or ω-dependent choice of functions. Although these axiom systems are increasing, they all have the same proof-theoretic strength as can be seen by setting up suitable models of ramified analysis (Kreisel in [51] and [28], 375-376).

Let us now consider the following questions: In how far do these axiomatizations allow the usual development of classical analysis and are there further genuine extensions of it by accepted principles of classical mathematics - say Zermelo-Fraenkel set theory ZFC - concerning only the above mentioned structures, i.e. numbers and functions but no uncountable cardinalities? As to the first question, the reader may convince himself that all of known analysis can be built up as a theory of sets of numbers within this axiomatic framework along the lines given in Hilbert-Bernays [16], Supplement IV. There three formal systems H,K,L are described in which likewise analysis can be developed:

H: ε-abstraction over numbers and functions
K: quantification over numbers and functions together with ι,λ-symbols
L: quantification over numbers and predicates

Our considerations below are based on the generalization of the formal system K to functionals of all finite types (functional language) omitting the ι- and λ-symbols which are eliminable therein as has been shown by Schütte [44]. This uniform generalization to finite types comes in very naturally with the ideas of functional interpretation and does not influence the proof-theoretic strength (see the last chapter).

The second question is answered by the following: Zermelo-Fraenkel set theory ZFC^- without the power set axiom is proof-theoretic equivalent to the axiom system of analysis (Kreisel [28], 375-376). Firstly, analysis is represented in the usual way in ZFC^-. Secondly, there is a ZFC^--model in analysis with axioms of choice by interpreting sets in a suitable way as hereditarily countable well-founded trees coded by number-theoretic functions and taking set equality as isomorphism between these trees and membership as isomorphism to direct subtrees. The axioms of choice can be dispensed with by using the above mentioned models of ramified analysis and the methods of Gödel and Cohen. Thus, analysis corresponds exactly to that part of known classical mathematics which makes no essential use of higher cardinalities, in other words it rests only on the denumerable.

The first attempts to show the correctness of the impredicative principles of analysis consisted in approximations and reinterpretations by theories which are built up from the outset in a correct and therefore consistent manner (see Schütte [46]). The first attack on full analysis however was made by Takeuti [60] in 1954 when he recognized that the consistency of analysis follows from the cut-elimination in a calculus of sequents for second (and higher) order logic (with comprehension). He then conjectured that Gentzen's Hauptsatz could be extended to these systems. A semantical criterion for this conjecture was given in 1960 by Schütte [45]. Based on these ideas, cut-elimination was then proved for second order logic by Tait [53] and Prawitz [34], and shortly afterwards for full simple type theory by Takahashi [57] and Prawitz [35]. Meanwhile these results were extended to extensionality and to intuitionistic versions of simple type theory (Prawitz [36]; Takahashi [58], [59]; Osswald [33]). However, the principles on which all these proofs of the Takeuti-conjecture are based are so inconstructive that they certainly are not a foundational basis. This situation has been improved by considering subsystems of analysis. So Takeuti [61] was able to reduce finitistically the consistency of Π_1^1-analysis (comprehension restricted to Π_1^1-formulae) to the well-ordering proof of his ordinal diagrams.

Gentzen's ideas have also been generalized via systems of natural deduction (see Prawitz [37], Kreisel [29]). The normal form theorem for the latter corresponds to the cut-elimination theorem in the calculus of sequents. For first order logics these purely existential statements can be proved constructively, thus yielding additionally the construction of the normal derivations, i.e. normalization theorems. These considerations can be extended with modifications or restrictions to first order theories (see Prawitz [37]). For second order logic, i.e. analysis, we get, at first, the classical existential normal form theorems by the cut-elimination proofs above. Recently Girard [10], Martin-Löf [32] and Prawitz [37] have obtained normalization theorems on the basis of some new ideas of Girard. Thus we have the same results for second order, but it must be stressed that the proofs use full impredicative comprehension in the domain of natural numbers and constitute therefore no reduction of the impredicative character.

The second of the two existing methods for far reaching consistency proofs, which will be studied subsequently, is functional interpretation. Gödel [12] introduced this method in 1958 for number theory by setting up an interpretation in the quantifier-free theory T of primitive recursive functionals of finite type. (The use of functionals in foundational investigations goes back to Hilbert [15].) To complete the consistency proof, one has to show the computability of the used terms; this can be done by means of transfinite induction up to ε_0 or by using higher abstract type-notions (see chapters VI, X). Gödel and others later on preferred the intensional view; our treatment below is extensional, which is of special interest in connection with stronger classical systems.

Gödel's functional interpretation was first applied to analysis by Kreisel [23] 1959, who proved classically the reduction to the so-called continuous (countable) functionals of finite type. Considerable progress in the constructive direction was then given by Spector [50] in his last paper. There he formulated - under the influence of the formalization of intuitionistic analysis by Kleene and Vesley [22] - the corresponding bar recursion principle to Brouwer's bar induction over numbers, generalized it to all finite types and gave the reduction of analysis to the functional calculus $T \cup BR$, i.e. T plus bar recursion in all finite types. Similar to the preceding case of number theory, the generalization to higher types constitutes a considerable strengthening of the combinatorial power. The problem remains to set up a foundation to these recursion principles, which means in the constructive case to bring them back in one way or an other to intuitionistic comprehensible dimensions.

Spector's result was improved by Howard [18] in that now the strongest form of ω-choice, namely dependent choice, was included. With respect to a foundation of the bar recursive functionals Kreisel stated in [27] p. 147, 151 (without detailed proofs) the following results:

a) The continuous functionals of finite type are a classical model for

T∪BR. Thus, at least classically, the reduction of analysis to T∪BR is meaningful.

b) The consistency of so far formalized intuitionistic analysis can be proved in classical Π_1^1-analysis. Thus only bar recursion for the types 0, 00, 000,... can be established by accepted intuitionistic principles, and it is (by Gödel's second consistency theorem) impossible to do so for type 0(00) and all higher types because this would imply the consistency of Π_1^1-analysis. Therefore to give an intuitionistic foundation to classical analysis one has to strengthen existent intuitionistic analysis in a considerable way. Our proof-theoretic experience lets us guess that for our purposes this strengthening must be by some far-reaching inductive process.

The essential part of result a) above was proved rigorously by Scarpellini [38] by means of a more manageable classical topological model. Meanwhile Tait [56] also published a classical treatment of T∪BR.

The most constructive foundation of T∪BR and thus of analysis so far proved was finished in the summer of 1970 by the author. It differs from the other works by giving a reduction to generalized inductive processes, thus following up Gentzen's ideas with respect to the principles (but not the methods) used. Before describing the main points thereof, to conclude the above explanations we have to mention still the following two things. Firstly, Scarpellini [40] after having seen our work was able to improve his topological model, thus yielding another proof by the same means (see chapter XIV) plus an additional axiom of choice; it may be mentioned that our model, which comes up directly from a computational analysis of the situation, is minimal (with respect to the method used, see below). Secondly, another sort of functional interpretation was recently developed by Girard [10]; but the handling of the corresponding functional calculus requires - as it stands now - full impredicative comprehension.

In the following, Gödel's functional interpretation and the known results about it are developed and rounded off systematically. As a new result e.g. it is shown that to every formula A a <u>construction statement</u> $\bigwedge x_1 \ldots x_m \bigvee y_1 \ldots y_n A_o$ (A_o quantifier-free) can effectively be exhibited which is classically equivalent to A, and which remains unchanged during the interpretation steps and is thus for provable A effectively satisfied in the corresponding functional calculus in which the functional interprétation can be carried out. These statements expressing the constructive content of the original propositions constitute the guide through the interpretation. In Hilbert's terminology: they are the real elements which when the functional interpretation can be carried out are provably not changed by the non-constructive ideal elements additionally used in classic mathematics. These construction statements constitute what may be called the combinatorial content of the theory and what can be identified with the notion of proof-theoretic strength.

On that occasion we may offer the following thesis: Each meaningful mathematical statement possesses a constructive content which can be effectively analysed by suitable recursion and uniformization principles. According to this thesis constructive reasoning is no essential restriction.

This thesis will be verified here for analysis, i.e. for that part of known classical mathematics which is built only upon the denumerable. This is done by extending Spector's reduction of analysis to the bar recursive functionals T∪BR to include the axiom of dependent choice (yielding another proof of Howard's result) and then by studying the computability of the functional operations involved there by intuitionistic methods, thus bringing together Hilbert's program and intuitionism. The main tools which will be used to that purpose are the relative abstract type notions "extensional computability" and "extensional computational equality". As already noted above those parts of T∪BR which can be proved to be computable by accepted intuitionistic means consist essentially in restricting bar recursion to the types 0, 00, 000,.... The main result given in these lectures is a constructive proof of the computability of full T∪BR. To do this one has to set up some new constructive principle, which will strengthen known intuitionistic analysis from below Π_1^1-analysis up to the full proof-theoretic strength of classical analysis. Thus our work gives significant contributions to both areas of foundation: proof theory and intuitionism.

First, a minimal model $\mathfrak{M}$, which is closed under the operations of T∪BR and choice sequences of all finite types, is built up by an intuitionistic generalized inductive definition. Roughly, $\mathfrak{M}$ is the full free-generated intuitionistic second number class. This extension of T∪BR is essential for our proof because closure with respect to choice sequences is used substantially to prove the computability. Besides intuitionistic arithmetical principles, this proof requires only the natural generalization of Brouwer's bar induction over natural numbers to intuitionistic accepted species; in our case to that part of $\mathfrak{M}$ which is computable. Therefore this generalized bar induction is, in our theory, the previously mentioned new principle which makes intuitionistic analysis exactly as strong as classical analysis. Concerning the constructiveness of that principle, an intuitive motivation is given which includes Brouwer's motivation for bar induction of type 0 as a special case. Based on all of these results, the author proposes to accept generalized bar induction over species as an intuitionistic principle. This implies intuitionistic foundations for classical analysis via the above described construction statements. As corollaries, we have consistency, and a constructive proof via gödelization that the $T\cup BR$-constructions correspond exactly to analysis.

It may be mentioned that the whole reasoning can also be interpreted classically. Transfinite induction over an initial segment of the third number class corresponds then to our use of generalized bar induction. But we stress the constructive viewpoint where generalized bar induction corresponds to transfinite induction over uniformly attainable initial segments.

Concluding, we point out the fact that our constructive proof for the countable $T\cup BR$ to be computable is essentially based on the nondenumerable intuitionistic species $\mathfrak{M}$. Whether this detour via the nondenumerable can be avoided constructively or not seems to be a fundamental question (compare Gödel [12], p. 281).

Using more technical terms the lectures are arranged in detail as follows.

In chapter I a formal system for classical analysis is set up within the language of functionals of finite type using equality $\overset{0}{=}$ between (natural) numbers as the only primitive predicate. For later use the axiomatization is given stepwise, beginning with intuitionistic predicate logic in Gödel's codification together with a short equivalence proof. Then tertium-non-datur, the equality axioms for $\overset{0}{=}$, the arithmetical (Peano) axioms and the defining equations of the primitive recursive functionals are added. The first theorems of the quantifier-free recursive functional theory - continued in chapter V - are given. - As analytical axioms we have at first extensionality for which several representations are treated. Then the possibilities and connections between comprehension, choice and higher inductions are discussed. The codification ends with the $(AC)^{0}$- respectively (ωAC)-analysis by adding choice with quantifier-free predicates

$(AC)^{\alpha,\beta}$-qf: $\bigwedge x^{\alpha} \bigvee y^{\beta} A(x,y) \rightarrow \bigvee z^{\beta\alpha} \bigwedge x^{\alpha} A(x,zx)$ A quantifier-free

and choice over type 0

$(AC)^{0,\beta}$: $\bigwedge x^{0} \bigvee y^{\beta} A(x,y) \rightarrow \bigvee z^{\beta 0} \bigwedge x^{0} A(x,zx)$

respectively the more extensive (dependent) ω-choice

$(\omega AC)^{\alpha}$: $\bigwedge x^{0}, y^{\alpha} \bigvee z^{\alpha} A(x,y,z) \rightarrow \bigvee x^{\alpha 0} \bigwedge y^{0} A(y,xy,x(y'))$.

Functional and quantifier contractions are then proved intuitionistically with the following quantifier-free rule of extensionality

$$(ER)\text{-qf}: \frac{A \rightarrow ru_1 \ldots u_m \overset{0}{=} su_1 \ldots u_m}{A \rightarrow t(r) \overset{0}{=} t(s)} \qquad u_1, \ldots, u_m \notin A, r, s; \ A \text{ quantifier-free.}$$

By the help of (AC)-qf each formula is shown to be equivalent to effectively constructible $\bigwedge\bigvee, \bigvee\bigwedge$ -formulae (construction statements).

<u>Chapter II</u> gives the elimination of extensionality by relativization in a form specially suitable for functional languages. This can be carried through for $(AC)^{0}$,(ωAC)-analysis if $(AC)^{\alpha,\beta}$-qf is restricted to the types $(0,\beta)$, $(0 \ldots 0,0)$. By what is described above this yields all of classical analysis. - In the following chapters there are proof-theoretic reductions of $(AC)^{0}$-, (ωAC)-analysis with extensionality restricted to (ER)-qf (denoted for short also by $(AC)^{0}$,(ωAC)-analysis). Altogether we therefore investigate $(AC)^{0}$,(ωAC)-analysis with extensionality restricted to (ER)-qf as well as $(AC)^{0}$,(ωAC)-analysis without any restrictions on extensionality but with the above mentioned restrictions on (AC)-qf. In addition to the already mentioned axioms of choice $(AC)^{0,\beta}$, $(AC)^{0..0,0}$-qf, $(\omega AC)^{\alpha}$ extensionality can be eliminated in the same manner also for extensional-unique choice, $(AC)^{\alpha,0}$ and comprehension $(C)^{\alpha}$.

The first step in functional interpretation, namely the translation of classical theories into intuitionistic approximated theories, is treated in <u>chapter III.</u> All existing translations are equivalent. The simplest translation $\neg\neg^{*}$ consists in doubly negating the formula and all its direct subformulae under an $\bigwedge$-quantifier. It is well known that classical arithmetic is transformed in this way into Heyting-arithmetic. But the $\neg\neg^{*}$-image of the $(AC)^{0}$,(ωAC)-analysis is only covered in Heyting-analysis plus (ER)-qf, Markov-principle

$(MP)^\alpha$: $\bigwedge x^\alpha \{A(x) \vee \neg A(x)\} \wedge \neg \bigwedge xA(x) \longrightarrow \bigvee x \neg A(x)$ and

$(\neg \bigwedge \neg)^\circ$: $\bigwedge x^\circ \neg \neg A(x) \longrightarrow \neg \neg \bigwedge xA(x)$ respectively plus the more extensive

$(\overset{\neg\neg}{\omega}AC)^\alpha$: $\bigwedge x^\circ, y^\alpha \neg \neg \bigvee z^\alpha A(x,y,z) \longrightarrow \neg \neg \bigvee x^{\alpha\circ} \bigwedge y^\circ A(y,xy,x(y'))$.

Heyting-analysis, (ER)-qf, (MP) and also

$(\overset{\bigvee}{\to})$: $\bigwedge x \{A(x) \vee \neg A(x)\} \wedge (\bigwedge xA(x) \to \bigvee yB(y)) \to \bigvee y(\bigwedge xA(x) \to B(y))$

have a primitive recursive functional interpretation while $(\neg \bigwedge \neg)^\circ$, $(\overset{\neg\neg}{\omega}AC)$
require bar recursion. It is shown that under certain conditions,
which are fulfilled here, formulae of the forms $\neg \bigvee \bigwedge ... \bigvee \bigwedge$, $\bigwedge \bigvee$ are not
changed by the $\neg \neg^*$-translation. The chapter closes with a discussion
of the role of comprehension and choice within these classical-intuitionistic translation operations.

Gödel's functional interpretation in the narrower sense, which attaches
to each formula F of the intuitionistic approximated theory a $\bigvee \bigwedge$-for-
mula F', is described in <u>chapter IV</u>. For provable formulae these $\bigvee$-func-
tionals will be later on effectively exhibited in a functional calculus.
It is shown that in setting up the functional interpretation in the narrower sense functional contractions can be managed ad libitum with the
help of primitive recursive functionals and (ER)-qf. Precise inspection
of the '-formation gives the deductive equivalence of F↔F' with (MP),
$(\overset{\bigvee}{\to})$ over Heyting-analysis plus (ER)-qf. The functional interpretation
embraces the no-counterexample-interpretation. Moreover the functional
calculus f, in which the interpretation can be given, contains the functional domain of the original classical theory $\mathcal{7}$, and only such functionals can be proved in $\bigwedge \bigvee$-form to exist in $\mathcal{7}$ which are also in f (plus
primitive recursive functionals). According to chapter I in our formal
systems of analysis each formula is provably equivalent to an $\bigwedge \bigvee$-for-
mula (construction statement) thus sharpening the no-counterexample-
interpretation to a direct constructive interpretation in terms of constructive functionals. Gödel's functional interpretation is not fully
determined by the logical connectives, but depends also on the prime
formulae; the enforcement of the functional interpretation therefore requires effective characteristic functionals for quantifier-free formulae,
i.e. the decidability of the prime formulae. Because we use only $\overset{\circ}{=}$ as
a primitive predicate this condition is automatically fulfilled here;
therefore we don't enter into the (more complicated) modification of
Gödel's functional interpretation given recently by Diller and Nahm
which avoids the dependence from the prime formulae and allows thus the
interpretation of theories with undecidable prime expressions (e.g.
higher intensional equalities). Under certain conditions, which are

fulfilled in our cases, functional interpretation implies also ω-consistency. In this connection the role of ω-rules in functional interpretation is discussed.

In chapter V the quantifier-free primitive recursive functional theory T - as far as we need it here - is developed in a suitable form. The proofs, which include a derivation of simultaneous recursion, are still intensional. These considerations will be continued in chapter VII where the bar recursive functionals are added.

The effective fulfillment of the functional interpretation (in the narrower sense) of Heyting-analysis plus (ER)-qf, (MP), $(\underset{\rightarrow}{\vee})$ in T is given briefly in chapter VI using extensional functional contractions. This yields a functional interpretation of classical arithmetic plus (ER)-qf, (AC)-qf in T. Therefore T has the same proof-theoretic strength as classical arithmetic with or without (ER)-qf, (AC)-qf and the theories between Heyting-arithmetic and Heyting-analysis plus (ER)-qf, (MP),$(\underset{\rightarrow}{\vee})$. This interpretation - (ER)-qf excepted - can be given fully intensionally by using simultaneous recursion.

Chapter VII begins with a summary of the known results in functional interpretation of analysis. Then the quantifier-free bar recursive functional theory $T\cup BR$ is evolved as far as necessary. $T\cup BR$ arises from T by addition of the bar recursion equations in a λ-free formulation and a constructive ω-rule (ω). (Such a rule is necessary because only equality of type 0 is present.) The derivations now demand extensionality with respect to sequences over arbitrary types. - An informal motivation clears up the sometimes complicated relations.

After these preparations the effective fulfillment of the functional interpretation (in the narrower sense) of Heyting-analysis plus (ER)-qf, (MP), $(\underset{\rightarrow}{\vee})$, $(\neg\bigwedge\neg)^{0}$, $(\neg\neg\omega AC)$ in $T\cup BR$ is given in chapter VIII by extending Spector's argument and using extensional functional contractions. This yields a functional interpretation of $(AC)^{0}$, (ωAC)-analysis with (ER)-qf in $T\cup BR$ which uses for $\bigwedge$-formulae in $(AC)^{0,\alpha}$, $(\omega AC)^{\alpha}$ exactly bar recursion of type α. As a consequence, for instance, we have an interpretation of Π^1_1-,Σ^1_1-analysis by bar recursion of type 0(00). - In this interpretation - (ER)-qf excepted - extensionality can (not be avoided as it stands but) be restricted to extensionality on sequences over arbitrary types by using simultaneous bar recursion.

In chapter IX some consequences of the above functional interpretation of analysis are compiled. - Δ-comprehension of sets of numbers over $\bigvee\bigwedge$-formulae, to which all other cases reduce, are interpreted by bar recursion of the type of the first quantifier. So bar recursion of type 00, 00 and 0(00) yields arithmetical, hyperarithmetical and Δ^1_2-comprehension respectively. - It is directed towards the functional interpretation of the doubly negated tertium-non-datur on numbers which is closely connected with comprehension of type 0 and depends on the prime formulae. Contrary to this the functional interpretation of axioms of choice does not depend on prime formulae and is therefore simpler. - Finally Church's thesis (ChT) is over Heyting's arithmetic not only incompatible with the tertium-non-datur for numbers but it is even possible to set up a special case of (ChT) whose negation is functional interpretable in T plus (ω) and bar recursion $(BR)^0$ of type 0. A precise analysis shows that (ChT) contradicts over Heyting-analysis plus (ω), $(BR)^0$ the consistency of the Markov-principle for numbers with 00-parameters and even $\bigwedge x^{00} \neg\neg \bigvee y^0 \longrightarrow \neg\neg \bigwedge x \bigvee y$.

To conclude functional interpretation of analysis, $T\cup BR$ must be recognized as a calculus of constructions. Therefore the aim of the following chapters is to prove the extensional computability of $T\cup BR$ or parts of it. As already mentioned the bar recursive functionals make it necessary to investigate extensions of $T\cup BR$. It suffices to consider functional domains which consist only of the operations of $T\cup BR$ and choice sequences in arbitrary types. In passing, a μ-operator is also considered. At the beginning of chapter X standard computability instructions are given. The proof of computability itself is based on the type notions "extensionally computable" and "extensional computational equal". Then it is proved that T plus iterated choice sequences over numbers is extensionally computable. It turns out that functional interpretation is completed by extensional computability. - A stopping computation is continuous. This implies directly the extensional computability of the mentioned μ-operator. Now bar recursion - for the present of type 0 - is included. These considerations can be extended to the types 0...0 using the Howard-Kreisel reduction of bar induction BI^{00}_D of type 00 to Brouwer's bar induction BI^0_D by means of strong continuity. If one has proved the extensional computability of such a more extensive functional domain, then the extensional computability of the corresponding $T\cup BR$-part follows. - What we have thus far obtained as intuitionistically established functional interpretations yields the arithmetical and hyperarithmetical comprehension but not the Π^1_1-,Δ^1_2-analysis. For this, as stressed above, stronger means are necessary than those presently formalized in intuitionistic analysis. - The above considerations generalize to bar recursion of higher types provided functional domains are available which are extensionally closed with respect to choice sequences over these types. Previously, iterated choice sequences over numbers yield extensionally all possibilities within these types. For higher types the situation is more complicated because it must be constructively made sure that choice sequences together with the T∪BR-operations do not lead out of the given functional domain.

Gödel conjectured in 1962 that this problem could be attached by intuitionistic generalized inductive definitions. Chapter XI treats generalized inductive definitions first from the classical standpoint on the basis of comprehension and transfinite approximation. Intuitionistically the situation is quite different because the previously allowed means are now available only to a very restricted extent. Nevertheless there are also promissing intuitionistic inductive processes. Ordinal numbers, continuous functionals of type 0(00) and BI_D^0 are treated. In doing so the common basis, namely the construction of countable trees and the possibility of free growing, is stressed.

The generalizations of these inductive processes to trees over species - proposed by the author as intuitionistically acceptable - are discussed in chapter XII. Bar induction BI_D generalized to species is motivated by uniformization as a Brouwer bar induction BI_D^0 on a higher level of abstraction. What intuitionistically means, when one speaks of higher trees, is outlined. Finally continuous functionals are formed on such trees. - A discussion of the attempts to use these means in functional interpretation and the difficulties which are bound up with follows. From this the author draws the conclusion of a more metamathematical procedure.

In chapter XIII intuitionistically, via generalized inductive definitions, a free growing minimal functional domain $\mathfrak{M}$ is built up which is closed under all T∪BR-operations and choice sequences in arbitrary types. The extensional computability of $\mathfrak{M}$ is proved intuitionistically by bar induction BI_D generalized to subspecies of $\mathfrak{M}$. By this T∪BR has also been proved to be extensionally computable and thus the functional interpretation together with all its foundational consequences has been settled for $(AC)^0$,(ωAC)-analysis constructively in the sense of the means employed.

<u>Chapter XIV</u> - added in the spring of 1971 - gives additional information on the type of arguments of the proceeding chapters and the principles used there. First it is shown by arithmetization of the "local" discussion that analysis is proof-theoretic exactly as strong as $T \cup BR$ or the second order system based on Heyting-arithmetic plus T and the rule of bar induction generalized to species. Our full "global" proof arises by adding a constructive ω-rule expressing the uniformity governing the "local" arguments. - Then various forms of bar induction generalized to species are discussed. Particularly, it is shown that for these generalizations the rules are deductively equivalent to the implications. The same applies to the Markov-principle in a suitable formulation. - Our treatment is also characterized by the fact that these generalized versions of bar induction to species have a functional interpretation (in the narrower sense) in $T \cup BR$. The proof thereof can also be read as a certain analysis of these generalizations. - Then some attention is given to the intuitionistic related domain $\mathfrak{L}$ whose functional interpretation (in the narrower sense) can be given in $T \cup BR$ or an extension of it. $\mathfrak{L}$ possesses the well-known "intuitionistic" properties of $\vee, \bigvee$ but does not include comprehension; however $\mathfrak{L}$ can provably be extended in a consistent manner by comprehension of type 0 having only parameters of type 0 (higher ordered parameters would yield higher typed comprehensions because (AC) is in $\mathfrak{L}$). - It follows from the last results that on the general intuitionistic basis which is accepted today generalized bar induction does not imply comprehension; also the opposite can be proved, so that comprehension and generalized bar induction are so far constructively incomparable. Therefore also the same must be said about the two known reductions of classical analysis: the one via Gentzen's methods using full impredicative comprehension and our treatment via Gödel functional interpretation using generalized bar induction. But in the author's opinion the reduction to inductive processes given here is more satisfactory.

The literature is listed at the end in <u>chapter XV</u>.

I. A formal system of classical analysis

Our investigations are based on the functional language. In the following formal system (Kodifikat in the sense of H.A. Schmidt [41], [42]) classical analysis can be developed along the lines given in Hilbert-Bernays [16], Supplement IV (see also Schütte [46], chapter XI).

A. Language

I. Types:

1. 0 is the type of the natural numbers (inclusive zero).
2. If α, β are types, so $\alpha(\beta)$ is the type of the functionals from type β-domain to type α-domain.

Type denotation: $\alpha, \beta, \gamma, \ldots$ (with indices)

Remark: Each type is of the form $0(\alpha_1)\ldots(\alpha_m)$. After Schönfinkel [43] all functional relations of finite type can be represented in that way.

II. Basic material:

1. For each type α free functional variables u^α, v^α, $w^\alpha, \ldots$ (with indices)
2. For each type α bound functional variables x^α, y^α, $z^\alpha, \ldots$ (with indices)
3. Signs for functional constants:
   a. 0 of type 0 (zero)
   b. ' of type 00 (successor)
   c. φ^α, ψ^β, $\chi^\gamma, \ldots$ (with indices) as signs for primitive recursive functionals of finite type. Each sign is attached to exactly one (finite) system of defining equations.
4. Equality between objects of type 0: $\overset{0}{=}$
5. Logical operators: $\wedge$ (and), $\vee$ (or), $\rightarrow$ (implies), $\bigvee$ (there exists), $\bigwedge$ (for all)
6. Parentheses

In metalanguage this basic material denotes itself. In case 1, 2 and 3c the signs represent also the whole sort of the type considered.

III. Definition of quasiterm and T-rank:

1. Each free and bound functional variable and each functional constant is a quasiterm of corresponding type and T-rank 0.
2. If $r^{\alpha(\beta)}$, s^β are quasiterms of type $\alpha(\beta)$, β with T-ranks a resp. b, then $r(s)$ is a quasiterm of type α and T-rank $\max(a,b)+1$.

Denotation for quasiterms: r^α, s^β, $t^\gamma, \ldots$ (with indices)

Quasiterms are of the forms: 0, $'(r)$, $\varphi(r_1)\ldots(r_m)$, $u(r_1)\ldots(r_m)$ or $x(r_1)\ldots(r_m)$. - Quasiterms without bound functional variables are called terms.

IV. Definition of quasiformula and F-rank:

1. $r^0 \overset{0}{=} s^0$ is a quasi(prime)formula of F-rank 0.
2. If A, B are quasiformulae with F-ranks a resp. b, then $(A)\wedge(B)$, $(A)\vee(B)$ and $(A)\rightarrow(B)$ are quasiformulae of F-rank max(a,b)+1.
3. If $A(x^\alpha)$ is a quasiformula of F-rank a containing the bound functional variable x^α not only within the scope of quantifiers $\bigvee, \bigwedge$ then $\bigvee x^\alpha(A(x))$, $\bigwedge x^\alpha(A(x))$ are quasiformulae of F-rank a+1. The scope of the uttermost quantifiers $\bigvee, \bigwedge$ consists of all x^α in A which do not belong to another quantifier scope within A.

Denotation for quasiformulae: A, B, C,... (with indices)

Quasiformulae in which all bound functional variables belong to a quantifier scope are called formulae.

We make use of the familiar economy of parentheses (brackets to the right). Within an expression the type of a quasiterm is indicated - if necessary at all - only in one place. As usual we write $(r)'$, $r \overset{0}{\neq} s$ instead of $'(r)$, $\neg(r \overset{0}{=} s)$. - By r(s), A(s), A(B) we refer to a quasiterm s or a quasiformula B which may occur at certain places in r resp. A. The quasiterm resp. quasiformulae which arise in replacing s, B by t, C and crossing out quantifiers with empty scope are denoted by r(t), A(t), A(C). - With respect to $\bigvee, \bigwedge$ the brackets apply to the full quantifier scope. - $\bigwedge_i$ denotes a (finite) conjunction and $\bigvee_i$ a (finite) disjunction. - $A \vdash B$ means that the formula B can be proved within the deductive framework considered plus formula A as an additional axiom. - All these notions generalize in the natural way to n-tupels. - $\underline{r}$, $\underline{u}$, $\underline{x}$ denote n-tupels of quasiterms and free resp. bound functional variables.

V. Definitions:

1. Equivalence: $A \leftrightarrow B :\equiv (A \rightarrow B) \wedge (B \rightarrow A)$
2. Numerals: 0; $1 :\equiv 0'$; $2 :\equiv 1'$; ...
3. Higher equalities: $r^\alpha \overset{\alpha}{=} s^\alpha :\equiv \bigwedge x_1^{\beta_1},\ldots,x_m^{\beta_m}(rx_1\ldots x_m \overset{0}{=} sx_1\ldots x_m)$ for $\alpha \equiv 0\beta_m\ldots\beta_1$
4. Falsum: $\curlywedge :\equiv (0 \overset{0}{=} 1)$
5. Verum: $\curlyvee :\equiv (0 \overset{0}{=} 0)$
6. Negation: $\neg A :\equiv (A \rightarrow \curlywedge)$

B. Deduction frame (system of axioms and rules)

The deduction frame deals only with formulae.

I. Logical axioms and rules:

1. Intuitionistic predicate logic

Axioms:

(Taut): $A \vee A \to A$, $A \to A \wedge A$

(Add) : $A \to A \vee B$, $A \wedge B \to A$

(Perm): $A \vee B \to B \vee A$, $A \wedge B \to B \wedge A$

($\perp$) : $\perp \to A$

(Q) : $\bigwedge x^{\alpha} A(x) \to A(r^{\alpha})$, $A(r^{\alpha}) \to \bigvee x^{\alpha} A(x)$ (term r)

Rules:

(Mp): $\dfrac{A,\ A \to B}{B}$ (Exp): $\dfrac{A \wedge B \to C}{A \to B \to C}$ (Imp): $\dfrac{A \to B \to C}{A \wedge B \to C}$

(Syll): $\dfrac{A \to B,\ B \to C}{A \to C}$ (Sum): $\dfrac{A \to B}{C \vee A \to C \vee B}$

($\bigwedge$): $\dfrac{A \to B(u^{\alpha})}{A \to \bigwedge x^{\alpha} B(x)}$

($\bigvee$) : $\dfrac{A(u^{\alpha}) \to B}{\bigvee x^{\alpha} A(x) \to B}$

under the condition that u^{α} does not occur in the formulae below the line

Remark: This axiom system of intuitionistic predicate logic, which is especially suited for our considerations, was given by Gödel [12],286. - To prove the equivalence with one of the familiar formalizations it suffices to consider the propositional part. Obviously only intuitionistic laws are derivable (correctness). To show the completeness we refer to the $\perp$-codification of intuitionistic propositional logic in H.A. Schmidt [42], 373. Herein D5, D7 can be replaced by D2 and D27 by D25 according to Satz 52 (p. 206-207) and Satz 89, 90 (p. 288-289) thus yielding the following simple axiom system of Heyting's propositional logic:

Axioms:

D1: $A \to B \to A$ D2: $A \to B \to C \longrightarrow A \to B \longrightarrow A \to C$ D23: $A \wedge B \to A$

D24: $A \wedge B \to B$ D25: $A \to B \to A \wedge B$ D46: $A \to A \vee B$ D47: $B \to A \vee B$

D48: $B \to A \longrightarrow C \to A \longrightarrow B \vee C \to A$ $J5^{o}$: $\perp \to A$

Rule: (Mp)

With the exception of D2, D48 the proofs of these axioms in Gödel's axiom system are elementary.

(FR) $\dfrac{A \to B \to C,\ A \to B}{A \to C}$

Proof: $A \to B \to C$ premise

$A \wedge B \to C$ (Imp)

$B \wedge A \to C$ (Perm),(Syll)

(*) $B \to A \to C$ (Exp)

$A \to B$ premise

$A \to A \to C$ (*),(Syll)

$A \wedge A \to C$ (Imp)

$A \to C$ (Taut),(Syll)

(∨) $\dfrac{A \to C,\ B \to C}{A \vee B \to C}$

Proof: $A \to C$ premise

(*) $A \vee B \to C \vee B$ (Sum),(Perm),(Syll)

$B \to C$ premise

$C \vee B \to C \vee C$ (Sum)

$C \vee B \to C$ (Taut),(Syll)

$A \vee B \to C$ (*),(Syll)

Proof of D2: $D \equiv: ((A \to B \to C) \wedge (A \to B)) \wedge A$

$D \to A,\ D \to A \to B,\ D \to A \to B \to C$ (Add),(Perm),(Syll)

$D \to B,\ D \to B \to C$ (FR)

$D \to C$ (FR)

D2 by (Exp).

Proof of D48: $D \equiv: (B \to A) \wedge (C \to A)$

$D \to B \to A,\ D \to C \to A$ (Add),(Perm),(Syll)

$B \to D \to A,\ C \to D \to A$ analogous (*) in (FR)-proof

$B \vee C \to D \to A$ (∨)

$D \to B \vee C \to A$ analogous (*) in (FR)-proof

D48 by (Exp).

2. Tertium-non-datur

(Tnd) $A \vee \neg A$

Remark: According to Satz 123 in H.A. Schmidt [42], 357 together with the above equivalence proof the addition of (Tnd) gives full classical predicate logic of first order extended to the functional language of all finite types.

3. Equality axioms for type 0

(G1) $u \overset{0}{=} u$

(G2) $u \overset{\circ}{=} v \wedge w \overset{\circ}{=} v \rightarrow u \overset{\circ}{=} w$

(G3) $u \overset{\circ}{=} v \rightarrow r(u) \overset{\circ}{=} r(v)$

Remark: (G2), (G3) are - already in the intuitionistic framework - replaceable by (G4) $u \overset{\circ}{=} v \wedge A(u) \rightarrow A(v)$. While (G4)$\vdash$(G2), (G3) is straightforward with (G1), the reversal is proved by induction on the F-rank of A with the help of symmetry and transitivity of $\overset{\circ}{=}$ following from (G1), (G2).

II. Arithmetical axioms and rules:

1. Peano-axioms

(P1) $u' \overset{\circ}{\neq} 0$

(P2) $u' \overset{\circ}{=} v' \rightarrow u \overset{\circ}{=} v$

(CI) $\dfrac{A(0),\ A(u^{o}) \rightarrow A(u')}{A(u)}$ $\qquad$ (u^{o} not in $A(0)$)

Remark: Term substitution $\dfrac{A(u^{\alpha})}{A(t^{\alpha})}$ (u denotes all free variables u in A) follows from ($\bigwedge$), (Q). - The condition on (CI) is necessary. $A(v) \equiv:$ $u \overset{\circ}{=} v \rightarrow v = 0$ would be otherwise a counterexample for $v \equiv u$. - (CI) applied to $A(u^{o}) \equiv: A_1(0) \wedge \bigwedge x(A_1(x) \rightarrow A_1(x')) \rightarrow A_1(u)$ yields $A_1(0) \wedge \bigwedge x(A_1(x) \rightarrow A_1(x')) \rightarrow A_1(u)$. Thus (CI) is equivalent to the usual formalization of complete induction.

2. Defining equations of primitive recursive functionals

(R1) $\varphi u_1 \ldots u_m \overset{\alpha}{=} r$ $\qquad$ ($m > 0$) where r contains at most the variables $u_1, \ldots, u_m$ and the functional constants 0,' or previously introduced functional signs.

(R2) $\begin{cases} \varphi 0 \overset{\alpha}{=} \psi \\ \varphi u' = \chi(\varphi u^{o})u \end{cases}$ $\qquad$ with the same stipulations on ψ, χ as in (R1)

(R1), (R2) have some effect on the given axiom system; so (P1), (P2) are now provable. It is also well known that for quantifierfree formulae now (Tnd) can be derived. These intuitionistic proofs are briefly compiled in the following, emphasizing at the same time that these quantifierfree propositions can also be proved quantifierfree.

(δ) $\delta 0 \overset{\circ}{=} 0$, $\delta u' = u$ $\qquad$ (R2),(R1)

Proof of (P1): Induction on u

I. $0' \neq 0$ $\qquad$ def.

II. $u'' \overset{\circ}{=} 0 \rightarrow \delta(u'') = \delta 0$ $\qquad$ (G3)

$u'' \overset{\circ}{=} 0 \rightarrow u' = 0$ (δ)

Thus $u' \neq 0 \rightarrow u'' \neq 0$.

<u>Proof of (P2):</u>

$u' \overset{\circ}{=} v' \rightarrow \delta(u') = \delta(v')$ (G3)

$\rightarrow \quad u = v$ (δ)

(P3) $u \neq 0 \rightarrow u \overset{\circ}{=} (\delta u)'$

<u>Proof:</u> Induction on u

I. $0 \neq 0 \rightarrow 0 = 1$ def.,(G1)

$\rightarrow 0 = (\delta 0)'$ (λ)

II. $u' = (\delta u')'$ (δ),(G3)

Thus $(u \neq 0 \rightarrow u = (\delta u)') \rightarrow u' \neq 0 \rightarrow u' = (\delta u')'$.

(P4) $u \overset{\circ}{=} 0 \vee u \neq 0$

<u>Proof:</u> Induction on u

1. $0 = 0 \vee 0 \neq 0$ (G1)
2. $u = 0 \vee u \neq 0 \rightarrow u' = 0 \vee u' \neq 0$ (P1)

(I) $Iu \overset{\circ}{=} u$

(ν) $\nu uv \overset{\circ}{=} (uv)'$

(+) $+0 = I, \; +u' \overset{\circ\circ}{=} \nu(+u^{\circ})$

(P5) $+uv' \overset{\circ}{=} +u'v = (+uv)'$

<u>Proof:</u> $+u'v = \nu(+u)v = (+uv)'$

Proof of $+uv' = +u'v$ by induction on u:

I. $+0v' = v' = (+0v)' = +0'v$

II. $+uv' = +u'v \rightarrow +u'v' = (+uv')' = (+u'v)' = +u''v$

(P6) $+uv \overset{\circ}{=} +vu$

<u>Proof:</u> Induction on u

I. $+0v = +v0$ by induction on v:

1. $+00 = +00$
2. $+0v = +v0 \rightarrow +0v' = (+0v)' = (+v0)' = +v'0$ (P5)

II. $+uv = +vu \rightarrow +u'v = (+uv)' = (+vu)' = +vu'$ (P5)

(P7) $u = 0 \wedge v = 0 \leftrightarrow +uv = 0$

<u>Proof:</u>

1. $u = 0 \wedge v = 0 \rightarrow +uv = +00 = 0$
2. According to (P6) it suffices to show: $+uv = 0 \rightarrow u = 0$. Induction on u:

I. $+0v = 0 \rightarrow 0 = 0$

II. $+u'v = 0 \rightarrow (+uv)' = 0 \rightarrow \lambda \rightarrow u = 0$ (P5),(P1)

(ε) $\varepsilon uv \overset{0}{=} \delta(uv)$

($-$) $-0 = I,\ -u' \overset{00}{=} \varepsilon(-u^0)$
($-uv$ corresponds to the non-negative difference $v \dot{-} u$.)

<u>Definition:</u> $r^0 \leqslant s^0 \equiv: s \geqslant r \equiv: (-sr=0)$
$s < r \equiv: r > s \equiv: (-sr \neq 0)$

(P8) $u \leqslant v \vee v < u$ (P4)

(P9) $-u'v' = -uv$
<u>Proof:</u> Induction on u
I. $-0'v' = \delta(-0v') = \delta(v') = v = -0v$
II. $-u'v' = -uv \rightarrow -u''v' = \delta(-u'v') = \delta(-uv) = -u'v$

(P10) $u>v \rightarrow u=+v(-vu)$
<u>Proof:</u> Induction on v
I. $u>0 \rightarrow +0(-0u)=+0u=u$
II. $\{\underbrace{u>v}_{-vu\neq 0} \rightarrow u=+v(-vu)\} \wedge \underbrace{u>v'}_{-v'u\neq 0} \rightarrow -v'u=\delta(-vu)\neq 0$
$\rightarrow -vu\neq 0 \wedge -vu=(\delta(-vu))'=(-v'u)'$ (P3)
$\rightarrow u=+v(-vu)=+v(-v'u)'=+v'(-v'u)$ (P5)

(P11) $-u(+u'\cdots'v) \neq 0$. By (P6): $u < u'\cdots'$
<u>Proof:</u> Induction on u
I. $-0(+0'\cdots'v)=+0'\cdots'v=0 \rightarrow 0'\cdots'=0 \rightarrow \text{⅄}$ (P7),(P1)
II. $-u(+u'\cdots'v)\neq 0 \rightarrow -u'(+u'\cdots''v)=-u'(+u'\cdots'v)'=-u(+u'\cdots'v)\neq 0$ (P5),(P9)

(P12) $v\geqslant u \rightarrow v=+u(-uv)$
<u>Proof:</u> Induction on u
I. $v \geqslant 0 \rightarrow +0(-0v)=-0v=v$
II. $V \equiv: (\underbrace{v\geqslant u}_{-vu=0} \rightarrow v=+u(-uv))$
$-vu=0 \wedge V \wedge -vu'=0 \wedge -uv=0 \rightarrow v=+u(-uv)=+u0=u \wedge -vu'=0$ (P6)
$\rightarrow -uu'=0 \rightarrow \text{⅄}$ (P11),(P6)
$-vu=0 \wedge V \wedge -vu'=0 \rightarrow -uv\neq 0 \wedge -uv=(\delta(-uv))' \wedge v=+u(-uv)$ (P3)
$\rightarrow v=+u(\delta(-uv))'=+u'(-u'v)$ (P5)
$\underbrace{-vu\neq 0}_{u>v} \wedge -vu'=0 \rightarrow u=+v(-vu) \wedge -v(+v(-vu))'=0$ (P10)
$\rightarrow -v(+v'(-vu))=0 \rightarrow \text{⅄}$ (P5),(P11)
From the last two lines by (P4): $V \wedge \underbrace{-vu'=0}_{v\geqslant u'} \rightarrow v=+u'(-u'v)$

(P13) The rule $\dfrac{A(0,v^0),\ A(u^0,0),\ A(u,v) \rightarrow A(u',v')}{A(u,v)}$ $(u,v \notin A(0,0))$ is derivable.

Proof:

$B(w^0) :\equiv A(w,+vw)$, $C(w^0) :\equiv A(+uw,w)$ (w new)

(✱) B(w), C(w) by induction on w:

I. $B(0) \equiv A(0,+v0) \leftrightarrow A(0,v)$ (P6); C(0) analogous.

II. $B(w) \equiv A(w,+vw) \rightarrow A(w',(+vw)') \leftrightarrow A(w',+vw') \equiv B(w')$ premise,(P5);

for C analogous.

From (✱): $u>v \rightarrow A(+(-vu)v,v)$; $v \geq u \rightarrow A(u,+(-uv)u)$

$\rightarrow A(u,v)$ (P10),(P6) $\rightarrow A(u,v)$ (P12),(P6)

With (P8) now A(u,v).

(P14) $u \overset{0}{=} v \vee u \neq v$

Proof:

$A(u,v) :\equiv (u \overset{0}{=} v \vee u \neq v)$

A(u,0) by (P4). Likewise A(0,v). $A(u,v) \rightarrow A(u',v')$ according to (P2), (G3). A(u,v) now by (P13).

(P15) (Tnd) is intuitionistically provable for quantifierfree formulae.

Proof: Induction on F-rank together with (P14), term substitution and intuitionistic propositional logic.

With this there are intuitionistically also characteristic functionals for quantifierfree formulae.

(P16) For quantifierfree formulae all laws of classical propositional logic are intuitionistically derivable.

Proof: I,1; (P15), remark to (Tnd) in I,2.

Finally in presence of (I) (G1) and (G2) can be contracted to comparativity $u \overset{0}{=} v \wedge u = w \rightarrow v = w$.

III. Analytical axioms

1. Extensionality

We extend Spector [50] by including extensionality.

(E1) $u \overset{\sigma}{=} v \rightarrow r(u) \overset{0}{=} r(v)$

Remark: (E1) is intuitionistically deductive equivalent to (E2): $u \overset{\sigma}{=} v \wedge A(u) \rightarrow A(v)$ (induction on F-rank) and to each of the following rules of extensionality

$$\text{(ER)}\quad \frac{A \longrightarrow s^{\alpha} w_1^{\beta_1} \dots w_m^{\beta_m} \overset{o}{=} t w_1 \dots w_m}{A \longrightarrow r(s) \overset{o}{=} r(t)} \qquad\qquad \frac{A \longrightarrow s w_1 \dots w_m = t w_1 \dots w_m}{A \wedge B(s) \longrightarrow B(t)}$$

under the condition: $w_1, \dots, w_m \notin A, s, t$ and $\alpha \equiv o\beta_m \dots \beta_1$.

<u>Denotation:</u> A scheme for formulae or rules - denoted by (...) - is further classified by restricting the signs for quasiformulae therein to prenex quasiformulae of a certain kind. This restriction is indicated by joining to (...) the particular quantifier forms corresponding in left-to-right order to the signs for quasiformulae in (...). In addition "qf" means "quantifierfree". - For instance (ER)-qf denotes the rule (ER) with quantifierfree A.

Let us compare in the above intuitionistic framework (without (E1)) the equality $\overset{\alpha}{=}$ with the equality relation $u \overset{\alpha}{\cong} v \equiv: \bigwedge x^{o\alpha}(xu \overset{o}{=} xv)$.

(E3) $u \overset{\alpha}{\cong} v \rightarrow r(u) \overset{o}{=} r(v)$ is intuitionistically provable without (E1).
<u>Proof:</u> Let u, $w_1, \dots, w_m$ be the variables of r. By (R1) there is a functional s with $s w_1 \dots w_m u \overset{o}{=} r(u, w_1, \dots, w_m)$. (E3) now follows from $\bigwedge x(xu \overset{o}{=} xv) \longrightarrow s w_1 \dots w_m u \overset{o}{=} s w_1 \dots w_m v$.

From (E3) we have by specializing and induction on the F-rank:

(E4) $u \overset{\alpha}{\cong} v \longrightarrow u \overset{\alpha}{=} v$ (E5) $u \overset{\alpha}{\cong} v \wedge A(u) \longrightarrow A(v)$

Thus $\overset{\alpha}{\cong}$ is stronger than $\overset{\alpha}{=}$. Identification of $\overset{\alpha}{\cong}$ and $\overset{\alpha}{=}$ yields exactly (E1).

(E6) $u \overset{\alpha}{=} v \longrightarrow u \overset{\alpha}{\cong} v$ is intuitionistically deductive equivalent to (E2).
<u>Proof:</u>

I. $u \overset{\alpha}{=} v \wedge u \overset{\alpha}{\cong} u \longrightarrow u \overset{\alpha}{\cong} v$	(E2)
II. $u \overset{\alpha}{=} v \longrightarrow u \overset{\alpha}{\cong} v$	premise
$\longrightarrow A(u) \longrightarrow A(v)$	(E5)

The use of $\overset{\alpha}{=}$ is called <u>intensional</u> iff within the context considered for $\overset{\alpha}{=}$ only reflexivity and (E1), (E2) is used but not the $\overset{\alpha}{\cong}$-definition with $\bigwedge$.

<u>2. Comprehension, choice, higher inductions</u>

We begin with a list of the principles to be concerned.

$(C)^{\alpha}$: $\bigvee y^{o\alpha} \bigwedge x^{\alpha}(yx \overset{o}{=} 0 \leftrightarrow A(x))$ (comprehension)

$(AC)^{\alpha,\beta}$: $\bigwedge x^{\alpha} \bigvee y^{\beta} A(x,y) \longrightarrow \bigvee z^{\beta\alpha} \bigwedge x^{\alpha} A(x,zx)$ (choice)

$(DC)^{\alpha}$: $\bigwedge x^{\alpha} \bigvee y^{\alpha} A(x,y) \longrightarrow \bigwedge x^{\alpha} \bigvee z^{\alpha o}(z0 \overset{\alpha}{=} x \wedge \bigwedge z_1^{o} A(zz_1, z(z_1')))$

(dependent ω-choice)

$(\omega AC)^{\alpha}$: $\bigwedge x^{o}, y^{\alpha} \bigvee z^{\alpha} A(x,y,z) \longrightarrow \bigvee x^{\alpha o} \bigwedge y^{o} A(y,xy,x(y'))$ (ω-choice)

$(TI)^{\alpha}$: $\bigwedge x^{\alpha o} \bigvee y^{o} \neg(xy \succ x(y')) \longrightarrow \bigwedge x^{\alpha}(\bigwedge y^{\alpha}(x \succ y \rightarrow A(y)) \rightarrow A(x)) \longrightarrow \bigwedge x^{\alpha} A(x)$

(transfinite induction)

To formalize bar induction some notions on sequences are necessary.

Functional definition by cases

$$\varphi u_1..u_m \overset{\alpha}{=} \begin{cases} \varphi_1 u_1..u_m & \text{if } \psi u_1...u_m \overset{o}{=} 0 \\ \varphi_2 u_1..u_m & \text{if } \psi u_1...u_m \neq 0 \end{cases}$$

is obtained as follows: by (R2) $\chi 0 = \varphi_1$, $\chi u' = \varphi_2$; thus $\varphi u_1..u_m \overset{\alpha}{=} \chi(\psi u_1..u_m)u_1..u_m$.

Now

$$<>u^{\alpha o} v^{\alpha o} w_1^{o} w_2^{o} \overset{\alpha}{=} \begin{cases} uw_2 & \text{if} \quad w_1 > w_2 \\ vw_2 & \text{if} \quad w_1 \leq w_2 \end{cases}$$

$$\sigma u_1..u_n \overset{o}{=} 0 \qquad \text{(zero functional)}$$

Moreover the following abbreviations are used:

$$\overline{r,s} \equiv: <>r^{\alpha o}\sigma^{\alpha o}s^{o}$$

$$\overline{r,s}*t \equiv: <>(<>r^{\alpha o}(\lambda x^{o}t^{\alpha})s^{o})\sigma(s')$$

Remark: Occasionally the λ-operator is employed for short. The transcription to our λ-free functional usage is in these cases always evident.

$(BI)^{\alpha}_{\{D, M\}}$: $\bigwedge x^{\alpha o} \bigvee y^{o} A(\overline{x,y};y) \wedge \begin{Bmatrix} \bigwedge x^{\alpha o}, y^{o}(A(\overline{x,y};y) \vee \neg A(\overline{x,y};y)) \\ \bigwedge x^{\alpha o}, y^{o}, z^{\alpha}(A(\overline{x,y};y) \longrightarrow A(\overline{x,y}*z;y')) \end{Bmatrix} \wedge$

$\bigwedge x^{\alpha o}, y^{o}(A(\overline{x,y};y) \rightarrow B(\overline{x,y};y)) \wedge \bigwedge x^{\alpha o}, y^{o}(\bigwedge z^{\alpha} B(\overline{x,y}*z;y') \rightarrow B(\overline{x,y};y)) \longrightarrow$

$B(\sigma;0)$

(bar induction of type α with $\begin{cases}\text{decidability} \\ \text{monotonicity}\end{cases}$ condition)

$$(BI)^{\alpha}_{G\{{D \atop M}}: \ \bigwedge x^{\alpha o} \bigvee y^{o} A(x,y) \ \wedge \left\{ \begin{array}{l} \bigwedge x^{\alpha o}, y^{o}(A(x,y) \vee \neg A(x,y)) \\ \bigwedge x^{\alpha o}, y^{o}(A(x,y) \rightarrow A(x,y')) \end{array} \right\} \wedge$$

$$\bigwedge x^{\alpha o}, y^{o}(A(x,y) \rightarrow B(\overline{x,y};y)) \wedge \bigwedge x^{\alpha o}, y^{o}(\bigwedge z^{\alpha} B(\overline{x,y}*z;y') \rightarrow B(\overline{x,y};y)) \rightarrow B(\sigma;0)$$

(general bar induction of type α with $\left\{ \begin{array}{l} \text{decidability} \\ \text{monotonicity} \end{array} \right.$ condition)

Howard and Kreisel [19] investigated the intuitionistic and classical relations among these principles. Classically the following holds ([19], 350-353):

$(DC)^{\alpha}$, $(\omega AC)^{\alpha}$, $(TI)^{\alpha}$, $(BI)^{\alpha}_{D}$, $(BI)^{\alpha}_{M}$, $(BI)^{\alpha}_{GD}$, $(BI)^{\alpha}_{GM}$ are all deductive equivalent; further $(\omega AC)^{\alpha} \vdash (AC)^{o,\alpha} \vdash (C)^{o}$, $(AC)^{\alpha,\beta} \vdash (C)^{\alpha}$.

Remark: The considerations in [19] can be expressed directly in our formal systems reading $A(\overline{u,v};v^{o})$ as a statement Pc on the finite functional sequence $c \equiv: u0,\dots,u(v-1)$ (if $v=0$ then c is the empty sequence) of length $\tilde{c} \equiv: v$. The recursion axioms (1.1), (1.2) there can be proved here easily; concerning functional contractions see (1.1) below.

In the so far given axiomatic framework the wellordering of objects of a type can be proved according to Zermelo by suitable applications of extensionality and higher axioms of choice. The addition of higher axioms of choice therefore involves the addition of considerable parts of set theory, especially essential characterizations of higher cardinalities. Here we consider the possibilities of choice on the lowest level.

As new axioms we add to the previous axioms BI, BII, BIII1

$(AC)^{\alpha,\beta}$-qf: $\bigwedge x \bigvee y A(x,y) \longrightarrow \bigvee z \bigwedge x A(x,zx)$ (A quantifierfree)

and on the one hand $(AC)^{o,\alpha}$ and on the other hand the more extensive $(\omega AC)^{\alpha}$ for arbitrary finite types α,β. These two formal systems are called $(AC)^{o}$- resp. (ωAC)-analysis. By the preceding they cover comprehension $(C)^{o}$ of type 0; therefore classical analysis can be developed within them as a theory of sets of natural numbers.

Obviously the above codification can be shortened at several places. But the formulation given has the advantage that it applies without changes to subsystems which will be discussed in the following.

(1.1) (Functional and quantifier contraction)

$\phi,\ \psi_o,\ \psi_1$ (α,β)-code $\Longleftrightarrow$: $\bigwedge x^\alpha, y^\beta \{\psi_o(\phi xy) \overset{\alpha}{=} x \wedge \psi_1(\phi xy) \overset{\beta}{=} y\}$

The calculus of the primitive recursive functionals, i.e. the quantifier-free part of the formal systems of analysis plus (ER)-qf and term substitution, has the following properties:

(a) To each type α an intuitionistic intensional (α,α)-code $\phi^{(\alpha)}$, $\psi_o^{(\alpha)}$, $\psi_1^{(\alpha)}$ can be given.

(b) For all types α, β by (ER)-qf an intuitionistic extensional (α,β)-code $\phi^{(\alpha,\beta)}$, $\psi_o^{(\alpha,\beta)}$, $\psi_1^{(\alpha,\beta)}$ can be given; for α or $\beta \equiv 0$ this coding is intensional.

Iteration yields a contraction of finitely many functionals.- If the calculus of the primitive recursive functionals is extended to full predicate logic, then also the following holds:

(c) In each formula intuitionistically extensional a finite sequence of $\bigwedge$- or $\bigvee$-quantifiers of arbitrary types can be contracted by (ER)-qf to one $\bigwedge$- resp. $\bigvee$-quantifier in a provable equivalent manner. For a finite sequence of quantifiers of the same kind and the same type or type 0 this coding is intensional.

These simplification processes are in general connected with a rising of types.

Proofs:

(a): (1) $\varphi_1^{\alpha\alpha\alpha}uv \overset{\alpha}{=} u$ (2) $\varphi_2^{\alpha\alpha\alpha}uv \overset{\alpha}{=} v$

(3) $\varphi_3^{\alpha\alpha\alpha o}0 \overset{\alpha\alpha\alpha}{=} \varphi_1,\ \varphi_3 w' = \varphi_2$ (4) $\phi^{\alpha o\alpha\alpha}uvw \overset{\alpha}{=} \varphi_3 wuv$

(5) $\psi_o^{\alpha(\alpha o)}u_1^{\alpha o} \overset{\alpha}{=} u_1 0$ (6) $\psi_1^{\alpha(\alpha o)}u_1^{\alpha o} \overset{\alpha}{=} u_1 1$

Now intensionally:

$\psi_o(\phi uv) \overset{\alpha}{=} \phi uv0 = \varphi_3 0uv = \varphi_1 uv = u$ (5),(4),(3),(1)

$\psi_1(\phi uv) \overset{\alpha}{=} \phi uv1 = \varphi_3 1uv = \varphi_2 uv = v$ (6),(4),(3),(2)

(b): Starting with the types $\alpha \equiv 0\alpha_1 \ldots \alpha_m$, $\beta \equiv 0\beta_1 \ldots \beta_n$ we give an extensional one-to-one characterization of these objects in the type $\gamma :\equiv 0\alpha_1 \ldots \alpha_m \beta_1 \ldots \beta_n$. Then (a) gives the required (α,β)-code.

(1) $\phi_1^{\gamma\alpha} u^\alpha v_n^{\beta_n} \ldots v_1^{\beta_1} \overset{o}{=} u$ (2) $\phi_1^{-1\alpha\gamma} w^\gamma \overset{o}{=} w\sigma^{\beta_n} \ldots \sigma^{\beta_1}$

(3) $\phi_2^{\gamma\beta} v^\beta v_n^{\beta_n} \ldots v_1^{\beta_1} u_m^{\alpha_m} \ldots u_1^{\alpha_1} \overset{o}{=} v v_n \ldots v_1$ (4) $\phi_2^{-1\beta\gamma} w^\gamma v_n^{\beta_n} . v_1^{\beta_1} \overset{o}{=} w v_n .. v_1 \sigma^{\alpha_m} . \sigma^{\alpha_1}$

Intensionally: (5) $\phi_1^{-1}(\phi_1 u^\alpha) \overset{\alpha}{=} \phi_1 u\sigma \ldots \sigma \overset{\alpha}{=} u$ (2),(1)

$\phi_2^{-1}(\phi_2 v^\beta) v_n^{\beta_n} \ldots v_1^{\beta_1} \overset{o}{=} \phi_2 v v_n \ldots v_1 \sigma \ldots \sigma \overset{o}{=} v v_n \ldots v_1$ (4),(3)

Hence it follows with (ER)-qf: (6) $\phi_2^{-1}(\phi_2 v^\beta) \overset{\beta}{=} v$. For $\beta \equiv 0$ (ER)-qf is superfluous. - According to (5), (6) u^α, v^β are transformed extensionally biunique into the type γ by $\phi_1 u$, $\phi_2 v$.

(7) $\phi^{(\alpha,\beta)} u^\alpha v^\beta \overset{\gamma o}{=} \phi^{(\gamma)}(\phi_1 u)(\phi_2 v)$ (8) $\psi_o^{(\alpha,\beta)} u_1^{\gamma o} \overset{\alpha}{=} \phi_1^{-1}(\psi_o^{(\gamma)} u_1)$

(9) $\psi_1^{(\alpha,\beta)} u_1^{\gamma o} \overset{\beta}{=} \phi_2^{-1}(\psi_1^{(\gamma)} u_1)$

Now the following holds:

$\psi_o^{(\alpha,\beta)}(\phi^{(\alpha,\beta)} u^\alpha v^\beta) \overset{\alpha}{=} \phi_1^{-1}\{\psi_o^{(\gamma)}(\phi^{(\gamma)}(\phi_1 u)(\phi_2 v))\} \overset{\alpha}{=} \phi_1^{-1}(\phi_1 u) = u$

(7),(8),(a),(E2), (5)

$\psi_1^{(\alpha,\beta)}(\phi^{(\alpha,\beta)} u^\alpha v^\beta) \overset{\beta}{=} \phi_2^{-1}\{\psi_1^{(\gamma)}(\phi^{(\gamma)}(\phi_1 u)(\phi_2 v))\} \overset{\beta}{=} \phi_2^{-1}(\phi_2 v) = v$

(7),(9),(a),(E2), (6)

(c): It suffices to give the proof for two uttermost quantifiers. From this by iteration and induction on the F-rank intuitionistically the full assertion (c) is obtained.

From (b) follows with (ER)-qf $A(u^\alpha, v^\beta) \longleftrightarrow A(\psi_o^{(\alpha,\beta)}(\phi^{(\alpha,\beta)} uv),$ $\psi_1^{(\alpha,\beta)}(\phi^{(\alpha,\beta)} uv))$; consequently $\bigvee\!\!\!\!\bigwedge x^\alpha, y^\beta A(x,y) \longleftrightarrow \bigvee\!\!\!\!\bigwedge z^{\gamma o} A(\psi_o^{(\alpha,\beta)} z, \psi_1^{(\alpha,\beta)} z)$ where γ comes from α, β as in (b).

For $\alpha \equiv \beta$ or $\alpha \equiv 0$ or $\beta \equiv 0$ the same can be done intensionally by (a) resp. (b).

Remark: (ER)-qf is only used for empty A. - In a similar way further types can be contracted intensionally; e.g. $\alpha \equiv 0\alpha_1 \ldots \alpha_m$, $\beta \equiv 0\alpha_1 \ldots \alpha_m \beta_1 \ldots \beta_n$ is intensionally codable in the type $\beta 0$. - In some cases also the type of the contraction can be reduced further; e.g. u^o, $v^{o\alpha_1 \ldots \alpha_m}$ is coded extensionally in $\phi uv = \lambda x_m^{\alpha_m} .. x_1^{\alpha_1} 2^u 3^{v x_m \ldots x_1}$, i.e. in the type of v. Also the type $\alpha\beta 0$ is already represented extensionally in the type $\alpha\beta$:

$$\varphi u^{o} v^{\beta} \stackrel{\alpha}{=} \psi^{\alpha\beta}(\lambda x_{n}^{\beta_n} \ldots x_{1}^{\beta_1} 2 u_3 v x_n \ldots x_1), \quad \beta \equiv 0\beta_1 \ldots \beta_n$$

with $\psi w^{\beta} \stackrel{\alpha}{=}: \varphi(\exp(0, w\sigma \ldots \sigma))(\lambda x_n \ldots x_1 \exp(1, w x_n \ldots x_1))$

Analogous each type 0 argument can be extensionally absorbed by arbitrary other arguments and each type α argument by other type α arguments. - These aspects will not be pursued further.

(1.2) With classical logic and (AC)-qf each formula is provably equivalent to a prenex formula with prefix $\bigwedge\bigvee, \bigvee\bigwedge$ (construction statement).- This simplification process is in general connected with rising of types.

<u>Proof:</u>
I. for $\bigvee\bigwedge$: It suffices to prove the theorem for prenex formulae A. Complete induction on the length l of the prefix.

1. l=0: trivial
2. l≠0:

case 1: $A \equiv \bigvee x B(x)$

$A \leftrightarrow \bigvee x \bigvee y \bigwedge z C(x,y,z)$ C quantifierfree   (ind.hyp. for B)

By (1.1) now $\bigvee$-contraction.

case 2: $A \equiv \bigwedge x B(x)$

(AC)-qf means: (1) $\bigwedge x \bigvee y C(x,y) \leftrightarrow \bigvee z \bigwedge x C(x,zx)$

(2) $\bigvee x \bigwedge y D(x,y) \leftrightarrow \bigwedge z \bigvee x D(x,zx)$ with quantifierfree C, D

$A \leftrightarrow \bigwedge x \bigvee y \bigwedge z E(x,y,z)$ E quantifierfree   (ind.hyp. for B)

$\leftrightarrow \bigwedge x \bigwedge z_1 \bigvee y E(x,y,z_1 y)$ (2)

$\bigwedge$-contraction (1.1) and (AC)-qf (1) give the desired result.

II. for $\bigwedge\bigvee$: By I: $\neg A \leftrightarrow \bigvee x \bigwedge y B(x,y)$ B quantifierfree

(1.1), (1.2) yield far reaching simplifications. For instace (AC)-$\bigvee$ reduces to (AC)-qf, (AC) to (AC)-$\bigwedge$ and (ωAC) to (ωAC)-$\bigwedge$. For (ωAC) the proof after the (1.2) application runs as follows:

$$\bigwedge x^{o}, y^{\alpha} \bigvee z^{\alpha} \bigvee x_1^{\beta} \bigwedge y_1 A(x,y,z,x_1,y_1) \longrightarrow \bigwedge x^{o}, y^{\alpha} \bigvee z_1^{\gamma o} \bigwedge y_1 A(x,y,\psi_o z_1, \psi_1 z_1, y_1) \quad (1.1)$$

$$\longrightarrow \bigwedge x^{o}, y_2^{\gamma o} \bigvee z_1^{\gamma o} \bigwedge y_1 A(x, \psi_o y_2, \psi_o z_1, \psi_1 z_1, y_1)$$

$$\longrightarrow \bigvee z_2^{\gamma oo} \bigwedge x^{o} \bigwedge y_1 A(x, \psi_o(z_2 x), \psi_o(z_2 x'), \psi_1(z_2 x'), y_1) \quad (\omega AC)^{\gamma o}\text{-}\bigwedge$$

$$\bigwedge x^{o}, y^{\alpha} \bigvee z^{\alpha} \bigvee x_1^{\beta} \bigwedge y_1 A(x,y,z,x_1,y_1) \longrightarrow \bigvee z^{\alpha o} \bigwedge x^{o} \bigvee x_1 \bigwedge y_1 A(x,zx,zx',x_1,y_1)$$

In the following we frequently use theorem (1.1) without mentioning it because its assumptions are relatively weak. Especially we simplify the functional interpretation by suitable contractions. Thus we take up as in Spector [50] the extensional standpoint which however is only based on (ER)-qf; this is possible because all functionals used are extensional. - Following the original intention of Gödel [12] the considerations below can also be largely expressed intensionally by a detailed enforcement in tupels. After Gödel [12], Tait [54] this is possible without restrictions for number theory. For analysis after Spector [50] the quantifierfree extensionality rule applied to sequences over arbitrary types is needed. In Howard [18] this is not stressed but used for the finite sequences c.

II. Elimination of extensionality

In Takeuti [60], Gandy [8] and Schütte [47] the axiom of extensionality is eliminated by relativization. Here we develop this idea in a form specially suitable for functional languages. For (ωAC), (C) and some special cases of (AC) the elimination is carried out. This yields a (E1)-elimination for $(AC)^{o}$- resp. (ωAC)-analysis with the axioms $(AC)^{\alpha,\beta}$-qf restricted to the types ($\alpha \equiv 0$, β arbitrary), ($\alpha \equiv 0\ldots 0$, $\beta \equiv 0$). We content us here with this result because by chapter I it covers classical analysis. - In the following chapters proof-theoretic reductions are given for $(AC)^{o}$- resp. (ωAC)-analysis with extensionality restricted to (ER)-qf. Thus we investigate $(AC)^{o}$, (ωAC)-analysis with the above restrictions on (AC)-qf as well as $(AC)^{o}$, (ωAC)-analysis with extensionality restricted to (ER)-qf.

Definition of Ex^{α}, $\overset{\alpha}{\underset{e}{=}}$ by induction on types:

I. $Ex^{0}(r^{0}) \equiv: \curlyvee$, $r^{0} \overset{0}{\underset{e}{=}} s^{0} \equiv: r \overset{0}{=} s$

II. $Ex^{\alpha}(r^{\alpha}) \equiv: \bigwedge x_1^{\alpha_1}, \ldots, x_m^{\alpha_m}; y_1^{\alpha_1}, \ldots, y_m^{\alpha_m} (\bigwedge_{i=1}^{m} x_i \overset{\alpha_i}{\underset{e}{=}} y_i \rightarrow r x_m \ldots x_1 \overset{0}{=} r y_m \ldots y_1)$

$r^{\alpha} \overset{\alpha}{\underset{e}{=}} s^{\alpha} \equiv: Ex^{\alpha}(r) \wedge Ex^{\alpha}(s) \wedge \bigwedge x_1^{\alpha_1}, \ldots, x_m^{\alpha_m} (\bigwedge_{i=1}^{m} Ex^{\alpha_i}(x_i) \rightarrow r x_m \ldots x_1 \overset{0}{=} s x_m \ldots x_1)$

for $\alpha \equiv 0\alpha_1 \ldots \alpha_m$

Definition of A_e:

1. $A_e \equiv: A$ for prime formulae A
2. $(A \overset{\wedge}{\underset{\rightarrow}{\vee}} B)_e \equiv: A_e \overset{\wedge}{\underset{\rightarrow}{\vee}} B_e$
3. $(\overset{\vee}{\wedge} x^{\alpha} A(x))_e \equiv: \overset{\vee}{\wedge} x^{\alpha} (Ex^{\alpha}(x) \overset{\wedge}{\rightarrow} A_e(x))$

For the defined connectives $\neg, \leftrightarrow$ we have: $(\neg A)_e \equiv \neg A_e$, $(A \leftrightarrow B)_e \equiv (A_e \leftrightarrow B_e)$.

Used are the abbreviations $Ex(\underline{r}) \equiv: \bigwedge_{i=1}^{m} Ex(r_i)$, $\underline{r} \underset{e}{=} \underline{s} \equiv: \bigwedge_{i=1}^{m} r_i \underset{e}{=} s_i$, where $\underline{r} \equiv (r_1, \ldots, r_m)$, $\underline{s} \equiv (s_1, \ldots, s_m)$.

The deductive framework for the following proofs is arithmetic $\mathcal{A}$ (section I, II of the deduction frame). The considerations are also largely intuitionistically valid (without (Tnd)); deviations are marked.

At first some direct consequences of the above definitions.

(2.1) $r \overset{\alpha}{\underset{e}{=}} s \longrightarrow s \overset{\alpha}{\underset{e}{=}} r$

(2.2) $r \overset{\alpha}{\underset{e}{=}} s \wedge s \overset{\alpha}{\underset{e}{=}} t \longrightarrow r \overset{\alpha}{\underset{e}{=}} t$

(2.3) $r \overset{\alpha}{\underset{e}{=}} r \longleftrightarrow Ex^{\alpha}(r)$, $\quad r \underset{e}{=} s \longleftrightarrow \bigwedge \underline{\underline{x}},\underline{\underline{y}}\{\underline{\underline{x}} \underset{e}{=} \underline{\underline{y}} \longrightarrow r\underline{\underline{x}} \overset{o}{=} s\underline{\underline{y}}\}$

(2.4) $r \overset{\alpha}{=} s \wedge Ex^{\alpha}(r) \longrightarrow Ex^{\alpha}(s)$

(2.5) $r \overset{\alpha}{=} s \wedge Ex^{\alpha}(r) \longrightarrow r \overset{\alpha}{\underset{e}{=}} s$

(2.6) $\bigwedge x^{\alpha} Ex^{\alpha}(x)$, $\overset{\alpha}{\underset{e}{=}}$-definition for all types is deductive equivalent

to (E1), $\overset{\alpha}{=}$-definition for all types.

<u>Proof:</u>

$\Rightarrow$: With $\varphi\underline{\underline{y}}u \overset{o}{=} r(u,\underline{\underline{y}})$ according to (R1), where $u,\underline{\underline{y}}$ are the variables of r.

$\Leftarrow$: Induction on types with iterated (E1)-application.

(2.7) $Ex(\underline{\underline{\varphi}}) \wedge \underline{\underline{u}} \underset{e}{=} \underline{\underline{v}} \longrightarrow r(\underline{\underline{u}}) \underset{e}{=} r(\underline{\underline{v}})$ where $\underline{\underline{\varphi}}$ are the functional constants,

$\underline{\underline{u}}$ the variables of $r(\underline{\underline{u}})$

<u>Proof:</u> Induction on T-rank l of r

I. l=0: case 1: $r(\underline{\underline{u}}) \equiv \varphi$ conclusion from premise with (2.3)

case 2: $r(\underline{\underline{u}}) \equiv u$ trivial

II. l≠0: $r(\underline{\underline{u}}) \equiv s(\underline{\underline{u}})(t(\underline{\underline{u}}))$

By induction hypothesis:

(1) $Ex(\underline{\underline{\varphi}}) \wedge \underline{\underline{u}} \underset{e}{=} \underline{\underline{v}} \longrightarrow s(\underline{\underline{u}}) \underset{e}{=} s(\underline{\underline{v}})$

(2) $Ex(\underline{\underline{\varphi}}) \wedge \underline{\underline{u}} \underset{e}{=} \underline{\underline{v}} \longrightarrow t(\underline{\underline{u}}) \underset{e}{=} t(\underline{\underline{v}})$

We have to show:

($*_1$) $Ex(\underline{\underline{\varphi}}) \wedge \underline{\underline{u}} \underset{e}{=} \underline{\underline{v}} \longrightarrow Ex(s(\underline{\underline{u}})(t(\underline{\underline{u}}))) \wedge Ex(s(\underline{\underline{v}})(t(\underline{\underline{v}})))$

($*_2$) $Ex(\underline{\underline{\varphi}}) \wedge \underline{\underline{u}} \underset{e}{=} \underline{\underline{v}} \longrightarrow Ex(\underline{\underline{w}}) \longrightarrow s(\underline{\underline{u}})(t(\underline{\underline{u}}))\underline{\underline{w}} \overset{o}{=} s(\underline{\underline{v}})(t(\underline{\underline{v}}))\underline{\underline{w}}$

<u>Ad ($*_1$):</u>

$Ex(\underline{\underline{\varphi}}) \wedge \underline{\underline{u}} \underset{e}{=} \underline{\underline{v}} \longrightarrow Ex(s(\underline{\underline{u}})) \wedge Ex(t(\underline{\underline{u}}))$ (1),(2),def.

$\longrightarrow \underline{\underline{w}}_1 \underset{e}{=} \underline{\underline{w}}_2 \longrightarrow s(\underline{\underline{u}})(t(\underline{\underline{u}}))\underline{\underline{w}}_1 \overset{o}{=} s(\underline{\underline{u}})(t(\underline{\underline{u}}))\underline{\underline{w}}_2$ (2.3),def.

$\longrightarrow Ex(s(\underline{\underline{u}})(t(\underline{\underline{u}})))$

Analogous: $Ex(\underline{\underline{\varphi}}) \wedge \underline{\underline{u}} \underset{e}{=} \underline{\underline{v}} \longrightarrow Ex(s(\underline{\underline{v}})(t(\underline{\underline{v}})))$

Ad $(*_2)$:

$Ex(\varphi) \wedge \underline{u} =_e \underline{v} \longrightarrow Ex(t(\underline{v})) \wedge Ex(\underline{w}) \longrightarrow s(\underline{u})(t(\underline{v}))\underline{w} \overset{\circ}{=} s(\underline{v})(t(\underline{v}))\underline{w}$ (1),def.

(3) $\longrightarrow Ex(\underline{w}) \longrightarrow s(\underline{u})(t(\underline{v}))\underline{w} \overset{\circ}{=} s(\underline{v})(t(\underline{v}))\underline{w}$ (2),def.

$Ex(\varphi) \wedge \underline{u} =_e \underline{v} \longrightarrow t(\underline{u}) =_e t(\underline{v}) \wedge Ex(\underline{w}) \longrightarrow s(\underline{u})(t(\underline{u}))\underline{w} \overset{\circ}{=} s(\underline{u})(t(\underline{v}))\underline{w}$ (1),def., (2.3)

$\longrightarrow Ex(\underline{w}) \longrightarrow s(\underline{u})(t(\underline{u}))\underline{w} \overset{\circ}{=} s(\underline{u})(t(\underline{v}))\underline{w}$ (2)

$\longrightarrow Ex(\underline{w}) \longrightarrow s(\underline{u})(t(\underline{u}))\underline{w} \overset{\circ}{=} s(\underline{v})(t(\underline{v}))\underline{w}$ (3)

(2.8) For the primitive recursive functional constants φ: $Ex(\varphi)$.

Proof: by induction on the generation of φ

I. $Ex(0)$ by definition. $Ex(') \equiv \bigwedge x^0, y^0 (x=y \longrightarrow x'=y')$ by (G3).

II. case 1: φ according to (R1)

(1) $\varphi\underline{u} \overset{\circ}{=} r(\underline{u})$, where r contains only 0,' and preceding functional constants $\underline{\psi}$ and at most the variables $\underline{u}$.

$Ex(\underline{\psi}) \wedge \underline{u}_1 =_e \underline{v}_1 \wedge \underline{u}_2 =_e \underline{v}_2 \longrightarrow r(\underline{u}_1)\underline{u}_2 \overset{\circ}{=} r(\underline{v}_1)\underline{v}_2$ (2.7),def.,I

$\longrightarrow \varphi\underline{u}_1\underline{u}_2 \overset{\circ}{=} \varphi\underline{v}_1\underline{v}_2$ (1)

Since $Ex(\underline{\psi})$ by induction hypothesis, so $Ex(\varphi)$.

case 2: φ according to (R2)

(2) $\varphi 0 = \psi$, $\varphi u' = \chi(\varphi u)u^0$ with preceding functional constants ψ, χ

(3) $Ex(\psi)$, $Ex(\chi)$ by induction hypothesis. We have to show:

$A(u^0) \equiv: \bigwedge v, \underline{w}_1, \underline{w}_2 (u \overset{\circ}{=} v \wedge \underline{w}_1 =_e \underline{w}_2 \longrightarrow \varphi u \underline{w}_1 \overset{\circ}{=} \varphi v \underline{w}_2)$

$0 \overset{\circ}{=} v \wedge \underline{w}_1 =_e \underline{w}_2 \longrightarrow \varphi 0 \underline{w}_1 \overset{\circ}{=} \psi\underline{w}_1 = \psi\underline{w}_2 = \varphi 0 \underline{w}_2 = \varphi v \underline{w}_2$ (2),(3),(G3)

(4) Thus $A(0)$.

$A(u^0) \longrightarrow Ex(\varphi u)$

$\longrightarrow u' \overset{\circ}{=} v \wedge \underline{w}_1 =_e \underline{w}_2 \longrightarrow \varphi u' \underline{w}_1 \overset{\circ}{=} \chi(\varphi u)u\underline{w}_1 \overset{\circ}{=} \chi(\varphi u)u\underline{w}_2 \overset{\circ}{=} \varphi u' \underline{w}_2 \overset{\circ}{=} \varphi v \underline{w}_2$ (2),(3),(G3)

Thus $A(u^0) \longrightarrow A(u')$. With (4) by (CI) now $A(u^0)$.

(2.9) (a) $\underline{u} =_e \underline{v} \longrightarrow r(\underline{u}) =_e r(\underline{v})$

(b) $Ex(\underline{u}) \longrightarrow Ex(r)$ — where $\underline{u}$ are the variables of $r(\underline{u})$.

(c) $Ex(r^{0 \ldots 0})$

Proof: (2.7), (2.8), (2.3), (G3), def.

(2.10) $Ex(\underline{u}) \wedge v \underset{e}{=} w \longrightarrow A_e(v) \longrightarrow A_e(w)$ where $\underline{u}, v$ is the stock of variables of $A(v)$

Proof: Induction on F-rank of A. We prove only the induction basis where $A(v)$ is a prime formula $r(\underline{u},v) \overset{\circ}{=} s(\underline{u},v)$. The logical structure is settled as usual; the Ex-relativizations for quantifications come from the Ex-premises of the induction hypothesis.

$$Ex(\underline{u}) \wedge v \underset{e}{=} w \longrightarrow r(\underline{u},v) \overset{\circ}{=} r(\underline{u},w) \wedge s(\underline{u},v) \overset{\circ}{=} s(\underline{u},w) \qquad (2.9),(2.3)$$
$$\longrightarrow r(\underline{u},v) \overset{\circ}{=} s(\underline{u},v) \longrightarrow r(\underline{u},w) \overset{\circ}{=} s(\underline{u},w)$$

(2.11) $Ex(\underline{u}) \wedge Ex(v) \wedge v \overset{\alpha}{=} w \longrightarrow A_e(v) \longrightarrow A_e(w)$ where $\underline{u}, v$ are the variables of $A(v)$

Proof: (2.10),(2.5)

(2.12) (a)$^\alpha$ $(\bigwedge x^\alpha Ex(x))_e$

(b)$^\alpha$ $Ex(u) \wedge Ex(v) \wedge (u \overset{\alpha}{\underset{e}{=}} v)_e \longrightarrow u \underset{e}{=} v$

(c) $A_e \longleftrightarrow (A_e)_e$

Proof:

1. (a)$^\alpha \wedge$ (b)$^\alpha$ by type induction

I. $\alpha \equiv 0$: trivial

II. $\alpha \equiv 0\alpha_1 \ldots \alpha_m$: It is to be shown

(a)$^\alpha$ $Ex^\alpha(u) \longrightarrow (Ex(u))_e$, i.e. $\bigwedge \underline{x},\underline{y}(\underline{x} \underset{e}{=} \underline{y} \rightarrow u\underline{x} \overset{\circ}{=} u\underline{y}) \longrightarrow Ex(\underline{v}) \wedge Ex(\underline{w}) \wedge (\underline{v} \underset{e}{=} \underline{w})_e$

$$\longrightarrow u\underline{v} \overset{\circ}{=} u\underline{w}$$

This follows from the induction hypothesis (b)$^{\alpha_i}$ $(i=1,\ldots,m)$.

(b)$^\alpha$ $Ex(u) \wedge Ex(v) \wedge (Ex(u))_e \wedge (Ex(v))_e \wedge \bigwedge \underline{x}(Ex(\underline{x}) \wedge (Ex(\underline{x}))_e \rightarrow u\underline{x} \overset{\circ}{=} v\underline{x}) \longrightarrow$

$$Ex(u) \wedge Ex(v) \wedge \bigwedge \underline{x}(Ex(\underline{x}) \rightarrow u\underline{x} \overset{\circ}{=} v\underline{x})$$

This follows from the induction hypothesis (a)$^{\alpha_i}$ $(i=1,\ldots,m)$.

2. (a): $Ex(u) \longleftrightarrow (Ex(u))_e \wedge Ex(u)$ yields by induction on the F-rank $A_e \longleftrightarrow (A_e)_e$.

(2.13) (Deduction theorem for $\mathcal{A}$)

$A \vdash_{\mathcal{A}} B \Longleftrightarrow \vdash_{\mathcal{A}} A \rightarrow B$ for closed formulae A

Proof:

$\Leftarrow$: trivial

$\Rightarrow$: Demonstration as usual by putting A as a new implicational premise before each formula of the proof figure of B from A in $\mathcal{A}$. This figure of formulae can be completed to a proof figure of $A \to B$ in $\mathcal{A}$.

(2.14) $\vdash_{\mathcal{A}} A \Longrightarrow \vdash_{\mathcal{A}} Ex(\underline{u}) \to A_e$ where $\underline{u}$ indicates all free variables of A

Proof: Induction on the length of a proof of A

The $()_e$-formation does not change the main propositional structure. Therefore (2.14) follows for all propositional axioms, (G1) - (G3), (P1), (P2) and the defining equations of the functional constants from the instruction considered by simple logical transformations. The same applies also to the heredity of (2.14) for the rules (Sum), (Exp), (Imp). It still remains to consider (Q), ($\bigwedge$), ($\bigvee$), (Mp), (Syll) and (CI).

Ad (Q): Let $\underline{u}$ be the variables of A(r).

1. $Ex(\underline{u}) \to (Ex(r) \to A_e(r)) \to A_e(r)$ (2.9)

   $Ex(\underline{u}) \to \bigwedge x(Ex(x) \to A_e(x)) \to A_e(r)$ (Q)

   i.e. $Ex(\underline{u}) \to (\bigwedge xA(x) \to A(r))_e$

2. $Ex(\underline{u}) \to A_e(r) \to Ex(r) \wedge A_e(r)$ (2.9)

   $Ex(\underline{u}) \to A_e(r) \to \bigvee x(Ex(x) \wedge A_e(x))$ (Q)

   i.e. $Ex(\underline{u}) \to (A(r) \to \bigvee xA(x))_e$

Ad ($\bigwedge$),($\bigvee$):

1. Let $\underline{u}$ be the variables of $A \to \bigwedge xB(x)$.

$Ex(\underline{u}) \wedge Ex(v) \to A_e \to B_e(v)$ premise of ($\bigwedge$), ind.hyp.

$Ex(\underline{u}) \to A_e \to \bigwedge x(Ex(x) \to B_e(x))$ ($\bigwedge$), because v appears only at the indicated places.

i.e. $Ex(\underline{u}) \to (A \to \bigwedge xB(x))_e$

2. Let $\underline{u}$ be the variables of $\bigvee xA(x) \to B$.

$Ex(\underline{u}) \wedge Ex(v) \to A_e(v) \to B_e$ premise of ($\bigvee$), ind.hyp.

$Ex(\underline{u}) \to \bigvee x(Ex(x) \wedge A_e(x)) \to B_e$ ($\bigvee$), because v appears only at the indicated places.

i.e. $Ex(\underline{u}) \to (\bigvee xA(x) \to B)_e$

Ad (Mp),(Syll):

1. Let $\underline{u}$,$\underline{v}$ be the stocks of variables of A resp. B.

$Ex(\underline{u}) \to A_e$, $Ex(\underline{u}) \wedge Ex(\underline{v}) \to A_e \to B_e$ premises of (Mp), ind.hyp.

$Ex(\underline{\underline{u}}) \wedge Ex(\underline{\underline{v}}) \rightarrow B_e$ (Syll)

$Ex(\underline{\underline{0}}) \rightarrow Ex(\underline{\underline{v}}) \rightarrow B_e$ substitution, contraction

$Ex(\underline{\underline{v}}) \rightarrow B_e$ (2.8),(Mp)

2. (Syll) is treated analogous.

Ad (CI):

Let $\underline{\underline{u}}$ be the variables of A(O).

$Ex(\underline{\underline{u}}) \rightarrow A_e(O)$, $Ex(\underline{\underline{u}}) \wedge Ex(v^O) \rightarrow A_e(v) \rightarrow A_e(v')$ premises of (CI), ind.hyp.

$(Ex(\underline{\underline{u}}) \rightarrow A_e(v^O)) \rightarrow (Ex(\underline{\underline{u}}) \rightarrow A_e(v'))$ Ex^O-def.,(G1)

$Ex(\underline{\underline{u}}) \rightarrow A_e(w^O)$ (CI), because $v \notin A(O)$

$Ex(\underline{\underline{u}}) \wedge Ex(w^O) \rightarrow A_e(w)$

(2.15) (Elimination theorem)

M_1, M_2 be sets of closed formulae with:

(*) $\mathcal{A} \cup M_1 \vdash A_e$ for all $A \in M_1 \cup M_2$

$\mathcal{A} \cup M_1 \cup M_2 \vdash B \Rightarrow \mathcal{A} \cup M_1 \vdash Ex(\underline{\underline{u}}) \rightarrow B_e$ where $\underline{\underline{u}}$ are the variables of B

Proof:

$\mathcal{A} \cup M_1 \cup M_2 \vdash B \Leftrightarrow$ There are $A_1,..,A_m \in M_1; B_1,..,B_n \in M_2$ $\vdash_{\mathcal{A}} \bigwedge_{i=1}^{m} A_i \wedge \bigwedge_{j=1}^{n} B_j \rightarrow B$ (2.13)

$\Rightarrow$ There are $A_1,..,A_m \in M_1; B_1,..,B_n \in M_2$ $\vdash_{\mathcal{A}} Ex(\underline{\underline{u}}) \rightarrow \bigwedge_i (A_i)_e \wedge \bigwedge_j (B_j)_e \rightarrow B_e$ (2.14)

$\Rightarrow \mathcal{A} \cup M_1 \vdash Ex(\underline{\underline{u}}) \rightarrow B_e$ (*)

Remark:

1. (2.15) gives a translation of $\mathcal{A} \cup M_1 \cup M_2$ in $\mathcal{A} \cup M_1$. Especially for $B \equiv \lambda$ the consistency of $\mathcal{A} \cup M_1 \cup M_2$ thereby reduces to that of $\mathcal{A} \cup M_1$.

2. In a consistent $\mathcal{A} \cup M_1$ negations of M_2-formulae cannot be proved, because for $C_1,\dots,C_m \in M_2$ we have:

$$\mathcal{A} \cup M_1 \vdash \neg \bigwedge_{i=1}^{m} C_i \Rightarrow \mathcal{A} \cup M_1 \cup M_2 \vdash \lambda \Rightarrow \mathcal{A} \cup M_1 \vdash \lambda \qquad (2.15)$$

3. (2.15) can intuitionistically be weakened to $\neg\neg$.

(2.16) M_1, M_2 be sets of closed formulae with:

(*) $\mathcal{A} \cup M_1 \vdash A_e$ for all $A \varepsilon M_1 \cup M_2$

(**) $\mathcal{A} \cup M_1 \cup M_2 \vdash \bigwedge x^\alpha Ex(x)$ for all α

$\mathcal{A} \cup M_1 \cup M_2 \vdash B \Longleftrightarrow \mathcal{A} \cup M_1 \vdash Ex(\underline{u}) \rightarrow B_e$ where $\underline{u}$ are the variables of B

Proof:

$\Rightarrow$: (2.15)

$\Leftarrow$: (**) neutralizes the $()_e$-relativization.

Remark:

1.(2.16) gives an interpretation of $\mathcal{A} \cup M_1 \cup M_2$ in $\mathcal{A} \cup M_1$.

2. Also (2.16) can intuitionistically be weakened to $\neg\neg$.

Of (2.16) we now make the following application:

M_{11} be the set of the $\bigwedge$-closed $(AC)^{o,\alpha}$-qf, $(AC)^{o\ldots o,o}$-qf and $(AC)^{o,\alpha}$ for all α.

M_{12} be M_{11} plus the $\bigwedge$-closures of $(\omega AC)^\alpha$ for all α.

M_2 be the set of the $\bigwedge x^\alpha Ex(x)$ for all α.

By (2.6) $\mathcal{A} \cup M_{11} \cup M_2$, $\mathcal{A} \cup M_{12} \cup M_2$ constitute the $(AC)^o$- resp. (ωAC)-analysis with the restrictions on (AC)-qf described in the beginning.

The premise (**) of (2.16) is trivially fulfilled by M_2-definition. By (2.12)(a) also one part of the premise (*) of (2.16) holds, namely:

$A \varepsilon M_2 \Rightarrow \vdash_{\mathcal{A}} A_e \Rightarrow \mathcal{A} \cup M_{11/2} \vdash A_e$.

It remains to show:

$(*_1)$ $A \varepsilon M_{11} \Rightarrow \mathcal{A} \cup M_{11} \vdash A_e$

$(*_2)$ $A \varepsilon M_{12} \Rightarrow \mathcal{A} \cup M_{12} \vdash A_e$

(2.17) $(AC)^{o,\alpha} \vdash_{\mathcal{A}} (\bigwedge (AC)^{o,\alpha})_e$

Proof: We have to show

$$(AC)^{o,\alpha} \vdash_{\mathcal{A}} Ex(\underline{u}) \longrightarrow \bigwedge x^o \{Ex^o(x) \rightarrow \bigvee y^\alpha (Ex^\alpha(y) \wedge A_e(x,y))\} \longrightarrow \bigvee z^{\alpha o} \{Ex^{\alpha o}(z) \wedge \bigwedge x^o (Ex^o(x) \rightarrow A_e(x,zx))\}$$

where $\underline{u}$ are the free variables of A.

$\bigwedge x^{o} \bigvee y^{\alpha}(Ex^{\alpha}(y) \wedge A_e(x,y)) \rightarrow \bigvee z^{\alpha o} \bigwedge x^{o}(Ex^{\alpha}(zx) \wedge A_e(x,zx))$ $\quad (AC)^{o,\alpha}$

$\rightarrow \bigvee z^{\alpha o}\{Ex^{\alpha o}(z) \wedge \bigwedge x^{o} A_e(x,zx)\}$

because by (G3) $\bigwedge x^{o} Ex^{\alpha}(zx) \rightarrow Ex^{\alpha o}(z)$

(2.18) $(AC)^{o\ldots o,o}\text{-qf} \vdash_{\mathcal{A}} (\bigwedge (AC)^{o\ldots o,o}\text{-qf})_e$

Proof: It is to be shown

$(AC)^{o..o,o}\text{-qf} \vdash_{\mathcal{A}} Ex(\underline{u}) \rightarrow \bigwedge x^{o..o}\{Ex^{o..o}(x) \rightarrow \bigvee y^{o}(Ex^{o}(y) \wedge A_e(x,y))\} \rightarrow$

$\bigvee z^{o(o..o)}\{Ex(z) \wedge \bigwedge x^{o}(Ex(x) \rightarrow A_e(x,zx))\}$

where $\underline{u}$ are the free variables of the quantifierfree $A \equiv A_e$

$Ex(\underline{u}) \wedge \bigwedge x^{o..o} \bigvee y^{o} A(x,y) \rightarrow \bigwedge x^{o..o} \bigvee y^{o}\{A(x,y) \wedge \bigwedge z_1^{o} < y \neg A(x,z_1)\}$

Kleene [17],190,*149a; (P15)

(*) $\rightarrow \bigvee z^{o(o..o)} \bigwedge x^{o..o}\{A(x,zx) \wedge \bigwedge z_1^{o} < zx \neg A(x,z_1)\}$

$(AC)^{o..o,o}\text{-qf}$

$Ex(\underline{u}) \wedge \bigwedge x^{o..o}\{A(x,zx) \wedge \bigwedge z_1^{o} < zx \neg A(x,z_1)\} \wedge v =_e w^{o..o}$

$\rightarrow Ex(\underline{u}) \wedge A(v,zv) \wedge \bigwedge z_1^{o} < zv \neg A(v,z_1) \wedge A(w,zw) \wedge \bigwedge z_1^{o} < zw \neg A(w,z_1) \wedge v =_e w$

$\rightarrow A(v,zv) \wedge \bigwedge z_1^{o} < zv \neg A(w,z_1) \wedge A(w,zw) \wedge \bigwedge z_1^{o} < zw \neg A(v,z_1)$ $\quad$ (2.10),def.

$\rightarrow zv \geq zw \wedge zw \geq zv$ $\quad$ (P8)

$\rightarrow zv \overset{o}{=} zw$ $\quad$ (T29),(T31),(T28) in chapter V

With (*) this yields

$Ex(\underline{u}) \wedge \bigwedge x^{o..o} \bigvee y^{o} A(x,y) \rightarrow \bigvee z^{o(o..o)}\{Ex(z) \wedge \bigwedge x^{o} A(x,zx)\}$

Now (2.18) follows with (2.9)(c), def..

By (2.17),(2.18) we have proved $(*_1)$ since $(AC)^{o,\alpha}$-qf is covered by (2.17). - As $M_{11} \subset M_{12}$, so it suffices for $(*_2)$ to prove the following.

(2.19) $(\omega AC)^{\alpha} \vdash_{\mathcal{A}} (\omega AC)^{\alpha}_e$

Proof: We have to show

$(\omega AC)^{\alpha} \vdash_{\mathcal{A}} Ex(\underline{u}) \rightarrow \bigwedge x^{o}, y^{\alpha}\{Ex^{o}(x) \wedge Ex(y) \rightarrow \bigvee z^{\alpha}(Ex(z) \wedge A_e(x,y,z))\} \rightarrow$

$\bigvee x^{\alpha o}\{Ex(x) \wedge \bigwedge y^{o}(Ex^{o}(y) \rightarrow A_e(y,xy,x(y')))\}$

where $\underline{u}$ are the free variables of A

(1) $Ex(\underline{u}) \wedge \bigwedge x^{o},y^{\alpha}\{Ex(y) \rightarrow \bigvee z^{\alpha}(Ex(z) \wedge A_e(x,y,z))\}$

$\longrightarrow Ex(u) \wedge \bigwedge x^{o},y^{\alpha}\bigvee z^{\alpha}\{Ex(y) \rightarrow Ex(z) \wedge A_e(x,y,z)\}$

$\longrightarrow \bigwedge x^{o},y^{\alpha}\bigvee z^{\alpha}\{(x\overset{o}{=}0 \rightarrow Ex(z) \wedge A_e(x,\sigma,z)) \wedge (x\overset{o}{\neq}0 \rightarrow Ex(y) \rightarrow Ex(z) \wedge$

$\wedge A_e(x,y,z))\}$ (2.8),(P4)

(2) $\longrightarrow \bigvee x^{\alpha o}\bigwedge y^{o}\{(y\overset{o}{=}0 \rightarrow Ex^{\alpha}(x(y')) \wedge A_e(y,\sigma,x(y'))) \wedge$

$(y\overset{o}{\neq}0 \rightarrow Ex^{\alpha}(xy) \rightarrow Ex^{\alpha}(x(y')) \wedge A_e(y,xy,x(y')))\}$ $(\omega AC)^{\alpha}$

(R1), (R2) give the functional relations:

(3) $\varphi_1 u^{o} v^{\alpha o} \overset{\alpha}{=} vu$

(4) $\varphi_2 u^{o}\ \alpha(\overset{o}{=}0)\ \sigma$ if $u\overset{o}{=}0$, $\varphi_2 u^{o} = \varphi_1 u$ otherwise

(5) $\psi v^{\alpha o} u^{o} \overset{\alpha}{=} \varphi_2 uv$

The following holds:

(6) $\psi v0 \overset{\alpha}{=} \varphi_2 0v = \sigma$ (5),(4)

(7) $\psi vu' \overset{\alpha}{=} \varphi_2 u'v = \varphi_1 u'v = v(u')$ (5),(4),(3)

By this we get from (2)

(8) $Ex(\underline{u}) \wedge \bigwedge x^{o},y^{\alpha}\{Ex(y) \rightarrow \bigvee z^{\alpha}(Ex(z) \wedge A_e(x,y,z))\} \longrightarrow$

$\bigvee x^{\alpha o}\bigwedge y^{o}\{Ex^{\alpha}(\psi x1) \wedge A_e(0,\psi x0,\psi x1) \wedge (y\overset{o}{\neq}0 \rightarrow Ex^{\alpha}(\psi xy) \rightarrow Ex^{\alpha}(\psi x(y'))$

$\wedge A_e(y,\psi xy,\psi x(y')))\}$ (G1),(6),(7),(P3),(2.4),(2.11),(G4),(2.8)

$B \equiv: Ex^{\alpha}(\psi v1) \wedge A_e(0,\psi v0,\psi v1) \wedge \bigwedge y^{o}(y\overset{o}{\neq}0 \rightarrow Ex^{\alpha}(\psi vy) \rightarrow Ex^{\alpha}(\psi v(y')) \wedge$

$A_e(y,\psi vy,\psi v(y')))$

$B \longrightarrow Ex^{\alpha}(\psi v0) \wedge A_e(0,\psi v0,\psi v1)$ (6),(2.8),(2.4)

$(B \rightarrow Ex^{\alpha}(\psi vw^{o}) \wedge A_e(w,\psi vw,\psi v(w'))) \longrightarrow B \longrightarrow Ex^{\alpha}(\psi v(w')) \wedge A_e(w',\psi v(w'),$

$\psi v(w''))$

((P4):$w=0 \vee w\neq 0$,(G4),(P1),B-def.: $y\equiv w,w'$)

Therefore according to (CI):

$B \rightarrow \bigwedge z^{o} Ex^{\alpha}(\psi vz) \wedge A_e(w,\psi vw,\psi v(w'))$

$B \rightarrow Ex^{\alpha o}(\psi v) \wedge A_e(w,\psi vw,\psi v(w'))$ (G3),def.

Together with (8) this gives

$Ex(\underline{u}) \wedge \bigwedge x^{o},y^{\alpha}\{Ex(y) \rightarrow \bigvee z^{\alpha}(Ex(z) \wedge A_e(x,y,z))\}$

$\rightarrow \bigvee x^{\alpha o}\bigwedge y^{o}\{Ex^{\alpha o}(\psi x) \wedge A_e(y,\psi xy,\psi x(y'))\}$

$\rightarrow \bigvee x^{\alpha o}\{Ex^{\alpha o}(x) \wedge \bigwedge y^{o} A_e(y,xy,x(y'))\}$

Remark:

Row (1) is without further argument intuitionistically valid only for

$\alpha \equiv 0\ldots0$ ((2.19)(c)). By $\Lambda\neg\neg \to \neg\neg\Lambda$ and the intuitionistic $(A \to \bigvee x B(x)) \to \neg\neg\bigvee x(A \to B(x))$ a $\neg\neg$-version can be obtained for which (E1)-elimination is possible by remark 2 of (2.16).

$(*_1)$, $(*_2)$ are entirely proved with (2.17) - (2.19). Therefore by (2.16) in the domains considered extensionality is eliminated via interpretation in the following sense:

$$\mathcal{A}\cup M_{11}\cup M_2 \vdash B \Longleftrightarrow \mathcal{A}\cup M_{11} \vdash Ex(\underline{\underline{u}}) \to B_e$$

$\underline{\underline{u}}$ are the variables of B.

$$\mathcal{A}\cup M_{12}\cup M_2 \vdash B \Longleftrightarrow \mathcal{A}\cup M_{12} \vdash Ex(\underline{\underline{u}}) \to B_e$$

In particular the consistency problem is thus $(B \equiv \curlywedge)$ reduced to the extensionality free part. According to remark 2 of (2.15) it follows from the consistency of the domains considered that the existence of intensional magnitudes is not provable within them.

Concluding the $()_e$-relativization is given in a similar way for $(C)^\alpha$ and two cases of choice which however are not further treated because a functional interpretation going beyond (ωAC)-analysis is not known today.

$$(AC!)^{\alpha,\beta} \quad \Lambda x^\alpha \bigvee y^\beta \{A(x,y) \wedge \Lambda z^\beta (A(x,z) \to y \underset{e}{=} z)\} \to \bigvee z^{\beta\alpha} \Lambda x^\alpha A(x,zx)$$

(Extensional unique choice)

(2.20) $(AC!)^{\alpha,\beta} \vdash_{\mathcal{A}} (\Lambda\, (AC!)^{\alpha,\beta})_e$

<u>Proof:</u> We have to show

$$(AC!)^{\alpha,\beta} \vdash_{\mathcal{A}} Ex(\underline{\underline{u}}) \to \Lambda x^\alpha \; (Ex(x) \to \bigvee y^\beta (Ex(y) \wedge A_e(x,y) \wedge \Lambda z^\beta (Ex(z) \wedge A_e(x,z) \to (y \underset{e}{=} z)_e))) \to \bigvee z^{\beta\alpha}\{Ex(z) \wedge \Lambda x^\alpha (Ex(x) \to A_e(x,zx))\}$$

where $\underline{\underline{u}}$ are the free variables of A

(1) $Ex(\underline{\underline{u}}) \wedge \Lambda x^\alpha \{Ex(x) \to \bigvee y^\beta (Ex(y) \wedge A_e(x,y) \wedge \Lambda z^\beta (Ex(z) \wedge A_e(x,z) \to (y \underset{e}{=} z)_e))\}$
$\to Ex(\underline{\underline{u}}) \wedge \Lambda x^\alpha \bigvee y^\beta \{\underbrace{(Ex(x) \to Ex(y) \wedge A_e(x,y) \wedge \Lambda z (Ex(z) \wedge A_e(x,z) \to y \underset{e}{=} z)) \wedge (\neg Ex(x) \to y \underset{e}{=} \sigma)}_{B(x,y)\;\equiv:}\}$ (2.12),(Tnd) for Ex(x),(2.8)

(2) $\to Ex(\underline{\underline{u}}) \wedge \Lambda x^\alpha \bigvee y^\beta \{B(x,y) \wedge \Lambda z^\beta (B(x,z) \to y \underset{e}{=} z)\}$ (Tnd) for Ex(x),(2.1), (2.2)

$$\mathrm{Ex}(\underline{u}) \wedge \Lambda x^{\alpha}\{\mathrm{Ex}(x) \to \mathrm{V}y^{\beta}(\mathrm{Ex}(y) \wedge A_e(x,y) \wedge \Lambda z^{\beta}(\mathrm{Ex}(z) \wedge A_e(x,z) \to (y \underset{e}{=} z)_e))\}$$

$$\to \mathrm{Ex}(\underline{u}) \wedge \mathrm{V}z^{\beta\alpha} \Lambda x^{\alpha} B(x,zx) \qquad (AC!)^{\alpha,\beta}$$

$$(3) \to \mathrm{Ex}(\underline{u}) \wedge \mathrm{V}z^{\beta\alpha} \Lambda x^{\alpha}\{\mathrm{Ex}(x) \to \mathrm{Ex}^{\beta}(zx) \wedge A_e(x,zx) \wedge \Lambda z_1^{\beta}(\mathrm{Ex}(z_1) \wedge A_e(x,z_1) \to zx \underset{e}{=} z_1)\} \qquad \text{(B-def.)}$$

$$\mathrm{Ex}(\underline{u}) \wedge \Lambda x^{\alpha}\{\mathrm{Ex}(x) \to \mathrm{Ex}^{\beta}(vx) \wedge A_e(x,vx) \wedge \Lambda z_1^{\beta}(\mathrm{Ex}(z_1) \wedge A_e(x,z_1) \to vx \underset{e}{=} z_1)\} \wedge w_1 \overset{\alpha}{\underset{e}{=}} w_2 \wedge \underline{w}_3 \underset{e}{=} \underline{w}_4 \to \mathrm{Ex}^{\alpha}(w_1) \wedge \mathrm{Ex}^{\alpha}(w_2) \wedge \mathrm{Ex}(\underline{u}) \wedge \mathrm{Ex}^{\beta}(vw_2) \wedge A_e(w_2,vw_2) \wedge (\mathrm{Ex}^{\beta}(vw_2) \wedge A_e(w_1,vw_2)$$

$$\to vw_1 \underset{e}{=} vw_2) \wedge w_1 \underset{e}{=} w_2 \wedge \underline{w}_3 \underset{e}{=} \underline{w}_4 \qquad \text{(def.)}$$

$$\to A_e(w_1,vw_2) \wedge (A_e(w_1,vw_2) \to vw_1 \underset{e}{=} vw_2) \wedge \mathrm{Ex}(vw_2) \wedge \underline{w}_3 \underset{e}{=} \underline{w}_4 \quad (2.10),(2.3)$$

$$\to \mathrm{Ex}(vw_2) \wedge vw_1 \underset{e}{=} vw_2 \wedge \underline{w}_3 \underset{e}{=} \underline{w}_4$$

$$\to vw_2\underline{w}_3 \overset{o}{=} vw_2\underline{w}_4 \wedge vw_1\underline{w}_3 \overset{o}{=} vw_2\underline{w}_3 \qquad \text{(def.)}$$

$$\to vw_1\underline{w}_3 \overset{o}{=} vw_2\underline{w}_4$$

Together with (3) this yields the assertion.

Remark:
Remark (2.19) applies here likewise. For the $\neg\neg$-version the (Tnd) used in (1), (2) passes into the intuitionistic $\neg\neg(A \vee \neg A)$. - The argumentation is preserved if one weakens $y \underset{e}{=} z$ in (AC!) to $\Lambda \underline{x}(\mathrm{Ex}(\underline{x}) \to y\underline{x} \overset{o}{=} z\underline{x})$.

(2.21) $(AC)^{\alpha,o} \underset{\mathcal{A}}{\vdash} (\Lambda (AC)^{\alpha,o})_e$

Proof:
One proof results by showing $(AC)^{\alpha,o} \Leftrightarrow (AC!)^{\alpha,o}$ with the least number principle and then going back to (2.20). Somehow directer is the proof which uses the least number principle after the $()_e$-relativization.

It must be shown

$$(AC)^{\alpha,o} \underset{\mathcal{A}}{\vdash} \mathrm{Ex}(\underline{u}) \to \Lambda x^{\alpha}\{\mathrm{Ex}(x) \to \mathrm{V}y^{o}(\mathrm{Ex}(y) \wedge A_e(x,y))\} \to \mathrm{V}z^{o\alpha}\{\mathrm{Ex}(z) \wedge \Lambda x^{\alpha}(\mathrm{Ex}(x) \to A_e(x,zx))\}$$

where $\underline{u}$ are the free variables of A

(1) $\mathrm{Ex}(\underline{u}) \wedge \Lambda x^{\alpha}\{\mathrm{Ex}(x) \to \mathrm{V}y^{o} A_e(x,y)\} \to \mathrm{Ex}(\underline{u}) \wedge \Lambda x^{\alpha} \mathrm{V}y^{o}\{\mathrm{Ex}(x) \to A_e(x,y)\}$

(2) $\to \mathrm{Ex}(\underline{u}) \wedge \Lambda x^{\alpha} \mathrm{V}y^{o}\{(\mathrm{Ex}(x) \to A_e(x,y)) \wedge \Lambda z^{o} < y \neg(\mathrm{Ex}(x) \to A_e(x,z))\}$

Kleene [17],190,*149°

(3) $\to \mathrm{Ex}(\underline{u}) \wedge \mathrm{V}z^{o\alpha} \Lambda x^{\alpha}\{(\mathrm{Ex}(x) \to A_e(x,zx)) \wedge \Lambda z_1^{o} < zx\ \neg(\mathrm{Ex}(x) \to A_e(x,z_1))\}$ $\qquad (AC)^{\alpha,o}$

$$Ex(\underline{u}) \wedge \bigwedge x^{\alpha}\{(Ex(x) \rightarrow A_e(x,v^{o\alpha}x)) \wedge \bigwedge z_1^o < vx \neg(Ex(x) \rightarrow A_e(x,z_1))\} \wedge w_1 \overset{\alpha}{\underset{e}{=}} w_2$$

$$\rightarrow Ex(\underline{u}) \wedge Ex(w_1) \wedge Ex(w_2) \wedge A_e(w_1,vw_1) \wedge \bigwedge z_1^o < vw_1 \neg(Ex(w_1) \rightarrow A_e(w_1,z_1))$$

$$\wedge A_e(w_2,vw_2) \wedge \bigwedge z_2^o < vw_2 \neg(Ex(w_2) \rightarrow A_e(w_2,z_2)) \wedge w_1 \underset{e}{=} w_2 \quad \text{(def.)}$$

$$\rightarrow A_e(w_1,vw_1) \wedge \bigwedge z_1^o < vw_1 \ \neg A_e(w_2,z_1) \wedge A_e(w_2,vw_2)$$

$$\wedge \bigwedge z_2^o < vw_2 \ \neg A_e(w_1,z_2) \quad \text{(2.10), def., (G1)}$$

$$\rightarrow vw_1 \geq vw_2 \wedge vw_2 \geq vw_1 \quad \text{(P8)}$$

$$\rightarrow vw_1 \overset{o}{=} vw_2 \quad \text{(T29),(T31),(T28) in chapter V}$$

From this with (3) the assertion follows.

Remark:

Remark (2.19) applies here likewise. In (2) now Kleene [17],190,*149b is used.

$$(2.22) \quad (C)^{\alpha} \underset{\mathcal{A}}{\vdash} (\bigwedge(C)^{\alpha})_e$$

Proof: We have to show

$$(C)^{\alpha} \underset{\mathcal{A}}{\vdash} Ex(\underline{u}) \rightarrow \bigvee z^{o\alpha}\{Ex(z) \wedge \bigwedge x^{\alpha}(Ex(x) \rightarrow (zx \overset{o}{=} 0 \leftrightarrow A_e(x)))\}$$

where $\underline{u}$ are the free variables of A

sg be the familiar functional of type 00 with:

$sg0 \overset{o}{=} 0, \quad sgu' = 1$

By (CI) (1) $sg\ u^o = 0 \vee sg\ u \overset{o}{=} 1$. (2) $u \overset{o}{=} 0 \leftrightarrow sg\ u = 0$ (G3),(P4),(P3)

(3) $\bigvee z^{o\alpha} \bigwedge x^{\alpha}(sg(zx) \overset{o}{=} 0 \leftrightarrow A_e(x))$ $\quad (C)^{\alpha}$,(2)

$$Ex(\underline{u}) \wedge \bigwedge x^{\alpha}(sg(v^{o\alpha}x) = 0 \leftrightarrow A_e(x)) \wedge w_1 \overset{\alpha}{\underset{e}{=}} w_2 \rightarrow (A_e(w_1) \leftrightarrow A_e(w_2)) \wedge$$

$$(sg(vw_1) = 0 \leftrightarrow sg(vw_2) = 0) \quad (2.10)$$

$$(4) \qquad \rightarrow sg(vw_1) \overset{o}{=} sg(vw_2) \quad (1)$$

By (R1): $\varphi v^{o\alpha} w^{\alpha} \overset{o}{=} sg(vw)$. (4) now means

$$Ex(\underline{u}) \wedge \bigwedge x^{\alpha}(sg(vx) = 0 \leftrightarrow A_e(x)) \rightarrow Ex^{o\alpha}(\varphi v) \wedge \bigwedge x^{\alpha}(\varphi vx \overset{o}{=} 0 \leftrightarrow A_e(x))$$

From this with (3) $Ex(\underline{u}) \rightarrow \bigvee z^{o\alpha}\{Ex(z) \wedge \bigwedge x^{\alpha}(zx \overset{o}{=} 0 \leftrightarrow A_e(x))\}$.

On account of (2.20) - (2.22) the given (E1)-elimination expands to the extensions of the above domains by $(AC!)^{\alpha,\beta}$, $(AC)^{\alpha,o}$ or $(C)^{\alpha}$.

Extensionality was eliminated in the domains considered by $()_e$- relativization. For $(AC)^o$, (ωAC)-analysis however $(AC)^{\alpha,\beta}$-qf was restricted to the types ($\alpha \equiv 0$, β arbitrary), ($\alpha \equiv 0\ldots 0$, $\beta \equiv 0$); presumably extensional choices for higher types cannot be ensured in this way without further arguments. - In the following chapters reductions are given for $(AC)^0$-, (ωAC) -analysis without these restrictions on (AC)-qf but with extensionality restricted to (ER)-qf (see end chapter I). For simplicity these formal systems are called again $(AC)^0$- resp. (ωAC)-analysis.

III. Translation of classical into intuitionistic approximated theories

By consistency proofs Hilbert meant considerations on the expressions of a formalized theory viewed as pure combinations of meaningless signs. Because the proofs catch the combinatorial content of the deduction frame for stronger theories they also reflect in a certain sense the constructive content of these theories. From the incompleteness theorems of Gödel follows not only that to prove the consistency even of number theory it is necessary to add to Hilbert's finitary mathematics of clear evidence on concrete objects certain abstract notions of higher level (by Gentzen parts of the constructive theory of ordinals) but that it is also important for the foundations of mathematics to use all informations available because there are consistent but e.g. ω-inconsistent calculi (Kleene [21],212).

Consistency proofs which from the first take their bearings from the constructive content were introduced by Gödel [12] in 1958 for number theory. Initially the classical theory is translated into an intuitionistic approximated theory. For the latter then the functional interpretation (in the narrower sense) is given by associating to each formula a $\bigvee\bigwedge$-formula. For provable formulae these $\bigvee$-magnitudes are now explicitly exhibited in a quantifierfree functional calculus. Thus the classical theory is thrown back to a functional calculus which is based on the notion of computable functional. To verify all propositions of this calculus one has to show the unambiguous computability of all number terms (resp. reduction of all terms) employed what requires constructive ordinals resp. corresponding higher abstract notions.

The so stated constructive content of classical propositions of course does not establish in general the correctness of these propositions in an intuitionistic sense. Proof theory is concerned to justify in this way classical mathematics which is so important for applications on account of its wealth and its elegance.

This chapter treats translations into intuitionistic approximated theories. - Shoenfield [49],214-222 has given a direct interpretation of classical number theory on $\neg, \vee, \bigwedge$-basis into the quantifierfree calculus of the primitive recursive functionals without mentioning an intuitionistic approximation. The same can also be done for analysis but does not alter the difficulties of the functional interpretation. It has the advantage of brevity but the disadvantage that it is not transparent what constructive interpretation is behind. Because it is just this we are interested in, we give a detailed representation.

The following table contains all today known classical-intuitionistic correspondences:

operation/ applied to	prime formulae P	$\neg A$	$A \wedge B$	$A \vee B$	$A \rightarrow B$
$^{-}$	$\neg\neg P$	$\neg\bar{A}$	$\bar{A} \wedge \bar{B}$	$\neg(\neg\bar{A} \wedge \neg\bar{B})$	$\neg(\bar{A} \wedge \neg\bar{B})$
^{o}	$\neg\neg P$	$\neg A^{o}$	$A^{o} \wedge B^{o}$	$\neg(\neg A^{o} \wedge \neg B^{o})$	$A^{o} \rightarrow B^{o}$
^{x}	$\neg\neg P$	$\neg\neg\neg A^{x}$	$\neg\neg(A^{x} \wedge B^{x})$	$\neg\neg(A^{x} \vee B^{x})$	$\neg\neg(A^{x} \rightarrow B^{x})$
*	P	$\neg A^{*}$	$A^{*} \wedge B^{*}$	$A^{*} \vee B^{*}$	$A^{*} \rightarrow B^{*}$

	$\bigwedge x A(x)$	$\bigvee x A(x)$	prefix
$^{-}$	$\bigwedge x \overline{A(x)}$	$\neg \bigwedge x \neg \overline{A(x)}$	-
^{o}	$\bigwedge x (A(x))^{o}$	$\neg \bigwedge x \neg (A(x))^{o}$	-
^{x}	$\neg\neg \bigwedge x (A(x))^{x}$	$\neg\neg \bigvee x (A(x))^{x}$	-
*	$\bigwedge x \neg\neg (A(x))^{*}$	$\bigvee x (A(x))^{*}$	$\neg\neg$

The translations $^{-}$,^{o} are given for number theory by Gödel [11] resp. Gentzen [9],532 and Bernays. Kleene [21],492-497 incorporated the pure logical part by double negationing of the prime formulae. Formulating the principle behind these translations the author met with the translations ^{x},$\neg\neg{}^{*}$; later on he observed that they were already stated in Kuroda [31].

For the following considerations the language may be extended by predicate variables as usual.

(3.1) For formulae A containing only the connectives $\neg, \wedge, \rightarrow, \bigwedge$ and only doubly negationed prime formulae intuitionistically holds: $\neg\neg A \leftrightarrow A$.
Proof: Induction on F-rank. Kleene [21],495/496.

(3.2) Intuitionistically: $\bar{A} \leftrightarrow A^{o}$
Proof: Induction on F-rank with $\neg(\bar{A} \wedge \neg\bar{B}) \leftrightarrow (\bar{A} \rightarrow \neg\neg\bar{B}) \leftrightarrow (\bar{A} \rightarrow \bar{B})$, (3.1).

(3.3) Intutionistically: $\bar{A} \leftrightarrow A^x$

Proof: Induction on F-rank with $\neg\neg\neg \leftrightarrow \neg$, $\neg\neg A \wedge \neg\neg B \leftrightarrow \neg\neg(A \wedge B)$, (3.1), $\neg\neg(A \vee B) \leftrightarrow \neg(\neg A \wedge \neg B)$, $\neg\neg(A \rightarrow B) \leftrightarrow (A \rightarrow \neg\neg B) \leftrightarrow \neg(A \wedge \neg B)$, $\neg\neg\bigwedge\neg\neg \leftrightarrow \bigwedge\neg\neg$, $\neg\neg\bigvee \leftrightarrow \neg\bigwedge\neg$.

(3.4) Intuitionistically: $\neg\neg A^* \leftrightarrow A^x$

Proof: Starting from the inside of A^x in each subformula containing only the connectives $\neg, \wedge, \vee, \rightarrow, \bigvee$ all double negations of the ^{x}-operation up to an outermost can be removed by the intuitionistic laws $\neg(\neg\neg A) \leftrightarrow \neg A$, $\neg\neg(\neg\neg A \overset{\wedge}{\underset{\rightarrow}{\vee}} \neg\neg B) \leftrightarrow \neg\neg(A \overset{\wedge}{\underset{\rightarrow}{\vee}} B)$, $\neg\neg\bigvee x \neg\neg A(x) \leftrightarrow \neg\neg\bigvee x A(x)$. Because we have for $\bigwedge$ intuitionistically only $\neg\neg\bigwedge\neg\neg \leftrightarrow \bigwedge\neg\neg$ at $\bigwedge$-quantifiers the inner $\neg\neg$ remain (according to the *-operation); the outer $\neg\neg$ are further drawn outside in the manner described previously.

Thus the above classical-intuitionistic translations are all equivalent. They rest on the $\neg\neg$-prefixing where (Tnd) passes into the intuitionistic $\neg\neg$(Tnd). Obviously on this basis the above table can be completed by some - but now trivial - modifications. Of all these possibilities the $\neg\neg^*$-version may be one of the most practical.

Below the following principles play a rôle:

$(MP)^\alpha$: $\bigwedge x^\alpha (A(x) \vee \neg A(x)) \wedge \neg\bigwedge x A(x) \longrightarrow \bigvee x \neg A(x)$ (so-called Markov-principle)

$(\underset{\rightarrow}{\bigvee})$: $\bigwedge x(A(x) \vee \neg A(x)) \wedge (\bigwedge x A(x) \rightarrow \bigvee y B(y)) \longrightarrow \bigvee y(\bigwedge x A(x) \rightarrow B(y))$ (" $\bigvee$ across $\rightarrow$ ")

(MP),$(\underset{\rightarrow}{\bigvee})$ are primitive recursive functional interpretable (see below) and their $\neg\neg$-versions are intuitionistic:

$$\neg\neg(MP): \bigwedge x(A(x) \vee \neg A(x)) \longrightarrow \bigwedge x \neg\neg A(x) \longrightarrow \bigwedge x A(x) \longrightarrow \neg\bigwedge x A(x) \longrightarrow \underbrace{\neg\bigwedge x \neg\neg A(x)}_{\neg\neg\bigvee x \neg}$$

$\neg\neg(\underset{\rightarrow}{\bigvee})$: Kleene [17],487

A quite different behavior have

$(\overset{\neg}{\bigwedge}\overset{\neg}{})^\alpha$: $\bigwedge x^\alpha \neg\neg A(x) \longrightarrow \neg\neg\bigwedge x A(x)$ (" $\neg\neg$ across $\bigwedge$ ")

$(\overset{\neg\neg}{\omega}AC)^\alpha$: $\bigwedge x^o, y^\alpha \neg\neg\bigvee z^\alpha A(x,y,z) \rightarrow \neg\neg\bigvee x^{\alpha o}\bigwedge y^o A(y,xy,x(y'))$ (" $\neg\neg$ across (ωAC)")

$(\neg\bigwedge\neg)$, $(\omega\neg\neg AC)$ are intuitionistically equivalent to their $\neg\neg$-versions. While $(\neg\bigwedge\neg)$-$\bigwedge$, $(\neg\bigwedge\neg)$-$\bigvee$, $(\omega\neg\neg AC)$-$\bigvee$ are still intuitionistic or follow intuitionistically with (MP) $(\omega\neg\neg AC)$-$\bigwedge$, $(\neg\bigwedge\neg)$-$\bigvee\bigwedge$ represent the proper problem for the functional interpretation below. $(\neg\bigwedge\neg)^{\alpha}$ is intuitionistically equivalent to $\neg\bigwedge x^{\alpha} \leftrightarrow \neg\bigwedge x\neg\neg$ or $\neg\neg(\neg\bigwedge x^{\alpha} \leftrightarrow \bigvee x\neg)$ and by $\neg\neg(A \vee \neg A)$, $\bigwedge x(A(x) \vee \neg A(x)) \rightarrow \bigwedge x\neg\neg A(x) \rightarrow \bigwedge xA(x)$ intuitionistically deductive equivalent to $\neg\neg\bigwedge x^{\alpha}(A(x) \vee \neg A(x))$. - $(\omega\neg\neg AC)$ embraces $(\neg\bigwedge\neg)^{o}$ for $A(x^{o},y,z) \equiv: B(x^{o})$.

(3.5) Intuitionistically plus $(\neg\bigwedge\neg)$: $\neg\neg A^{*} \leftrightarrow \neg\neg A$

Proof: From $(\neg\bigwedge\neg)$ follows $\neg\neg\bigwedge\neg\neg \leftrightarrow \bigwedge\neg\neg \leftrightarrow \neg\neg\bigwedge$. Herewith the $\neg\neg$-removal in the proof of (3.4) can be extended to $\bigwedge$. Altogether: $\neg\neg A^{*} \leftrightarrow A^{x} \leftrightarrow \neg\neg A$, (3.4).

(3.6) A is provable in classical (first order) predicate logic with equality iff $\neg\neg A^{*}$ is provable in intuitionistic (first order) predicate logic with equality.

Proof:

$\Leftarrow$: trivial

$\Rightarrow$: Induction on the length of a proof. By means of the intuitionistic laws $A \rightarrow \neg\neg A$, $\neg\neg(A \rightarrow B) \leftrightarrow (\neg\neg A \rightarrow \neg\neg B) \leftrightarrow (A \rightarrow \neg\neg B)$, $A \vee \neg\neg B \rightarrow \neg\neg(A \vee B)$ the $\neg\neg^{*}$-versions of the axioms and rules I1,3 of intuitionistic predicate logic with equality follow in each case easily from the instruction considered. (Tnd) passes by $\neg\neg^{*}$ into the intuitionistic $\neg\neg(A^{*} \vee \neg A^{*})$.

To extend (3.6) to theories over this logical framework we still have to consider the proper axioms. Because by (3.5) on the basis of certain $(\neg\bigwedge\neg)$-applications intuitionistically $A \rightarrow \neg\neg A \leftrightarrow \neg\neg A^{*}$ it is thus proved:

(3.7) Let T^{c} be a theory over classical (first order) predicate logic with equality and let T^{i} arise from T^{c} in replacing classical by intuitionistic predicate logic.

$T^{c} \vdash A \Longleftrightarrow T^{i}$ plus certain $(\neg\bigwedge\neg)$-cases (according to (3.5)) $\vdash \neg\neg A^{*}$

Especially T^{c} is equiconsistent to T^{i} plus these $(\neg\bigwedge\neg)$-cases.

Together with the results of chapter VI a constructive functional interpretation of certain $(\neg\bigwedge\neg)^{\alpha}$-cases therefore yields a constructive functional interpretation - and thus a constructive consistency proof - for all classical theories with constructive functional interpretable proper axioms for which the above considerations use only these $(\neg\bigwedge\neg)^{\alpha}$-cases. Because of this central position the functional interpretation of $(\neg\bigwedge\neg)$ constitutes a corresponding difficult task.

Remark:

1. From the intuitionistic equivalence of $(\neg\bigwedge\neg)$ with $\neg\neg\bigwedge(A \vee \neg A)$ and (3.7) by the deduction theorem follows:

$$T^c \vdash A \Leftrightarrow \text{There are } A_1,..,A_m \text{ with: } T^i \vdash \bigwedge_j \bigwedge \neg\neg\bigwedge(A_j \vee \neg A_j) \longrightarrow \neg\neg A^*$$

$$\Leftrightarrow \text{There are } A_1,..,A_m \text{ with: } T^i \vdash \neg\neg\bigwedge \bigwedge_j (A_j \vee \neg A_j) \longrightarrow \neg\neg A^* \quad (\neg\neg\bigwedge \rightarrow \bigwedge\neg\neg)$$

$$\Leftrightarrow \text{There exists B with: } T^i \vdash \neg\neg\bigwedge (B \vee \neg B) \longrightarrow \neg\neg A^* \quad (B \equiv: \bigwedge_j (A_j \vee \neg A_j), \ \neg\neg\bigwedge_j (A_j \vee \neg A_j))$$

Especially: $T^c \nvdash \perp \Leftrightarrow$ There is no B with: $T^i \vdash \neg\bigwedge (B \vee \neg B)$

The consistency of T^c thus means that the proper axioms of T^c do not refute intuitionistically (Tnd).

2. For a mere $\neg\neg$-prefixing the proof of (3.6) would be the same except for rule $(\bigwedge)$. At this place full $(\neg\bigwedge\neg)$ is needed. So the $\neg\neg^*$-translation has the advantage in transforming this into the proper axioms where by (3.5) only <u>certain</u> $(\neg\bigwedge\neg)$-cases are necessary.

Compared with the general considerations in (3.7) the needed $(\neg\bigwedge\neg)$-cases sometimes can be further reduced in concrete applications by utilizing special facts. We will do this now for our formal systems. For that purpose we use the following, in the given order increasing intuitionistic formal systems:

Heyting-arithmetic	$\equiv$:	arithmetical part (axioms I,II) without (Tnd)
$(AC)^o$-Heyting-analysis	$\equiv$:	$(AC)^o$-analysis without (Tnd)
(ωAC)-Heyting-analysis	$\equiv$:	(ωAC)-analysis without (Tnd)
Heyting-analysis	$\equiv$:	Heyting-arithmetic plus (ER)-qf, axiom of choice for all types

(3.8) Arithmetic $\vdash A \Leftrightarrow$ Heyting-arithmetic $\vdash \neg\neg A^*$ (Gödel [11])

<u>Proof:</u> The proof of (3.6) extends to this.

(3.9) Intuitionistically plus (E1): $\neg\neg(E1)^*$. Analogous for (ER),(ER)-qf.

Proof: $\underline{a}$-def., $\neg\neg\bigwedge\neg\neg \leftrightarrow \bigwedge\neg\neg$, (P16).

So the elimination of extensionality in chapter II is not necessary for the $\neg\neg^{*}$-translation but for the following functional interpretation (in the narrower sense).

(3.10) Intutionistically plus (AC)-qf, (MP): $\neg\neg((AC)\text{-qf})^{*}$

Proof:

$$\neg\neg((AC)\text{-qf})^{*} \equiv \neg\neg\{\bigwedge x\neg\neg\bigvee y A(x,y) \rightarrow \bigvee z \bigwedge x\neg\neg A(x,zx)\} \qquad \text{A quantifierfree}$$

$$\begin{aligned} \bigwedge x\neg\neg\bigvee y A(x,y) &\rightarrow \bigwedge x\neg\bigwedge y\neg A(x,y) \\ &\rightarrow \bigwedge x\bigvee y\neg\neg A(x,y) && \text{(MP),(P16)} \\ &\rightarrow \bigvee z\bigwedge x\neg\neg A(x,zx) && \text{(AC)-qf} \end{aligned}$$

(3.11) $(AC)^{o}$-analysis $\vdash A \Leftrightarrow (AC)^{o}$-Heyting-analysis plus (MP), $(\overset{\neg}{\bigwedge}\overset{\neg}{})^{o}\text{-}\bigvee\bigwedge \vdash \neg\neg A^{*}$

$\Leftrightarrow (AC)^{o}$-Heyting-analysis plus (MP), $(\underset{\rightarrow}{\bigvee})$, $(\overset{\neg}{\bigwedge}\overset{\neg}{})^{o}\text{-}\bigvee\bigwedge \vdash \neg\neg A^{*}$

$\Rightarrow$ Heyting-analysis plus (MP), $(\underset{\rightarrow}{\bigvee})$, $(\overset{\neg}{\bigwedge}\overset{\neg}{})^{o}\text{-}\bigvee\bigwedge \vdash \neg\neg A^{*}$

(Spector [50])

Proof:

Induction on the length of proof. On account of (3.8) - (3.10) we have still to consider $(AC)^{o,\alpha}$. By chapter I $(AC)^{o,\alpha}\text{-}\bigwedge$ suffices for it.

$$\neg\neg((AC)^{o,\alpha}\text{-}\bigwedge)^{*} \equiv \neg\neg\{\bigwedge x^{o}\neg\neg\bigvee y^{\alpha}\bigwedge z\neg\neg A(x,y,z) \rightarrow \bigvee z_1\bigwedge x\neg\neg\bigwedge z\neg\neg A(x,z_1x,z)\}$$

A quantifierfree

$$\leftrightarrow \neg\neg\{\bigwedge x\neg\neg\bigvee y\bigwedge z\neg\neg A(x,y,z) \rightarrow \bigwedge x\bigvee y\bigwedge z\neg\neg A(x,y,z)\}$$

$(\neg\neg\bigwedge\neg\neg \leftrightarrow \bigwedge\neg\neg, (AC)^{o,\alpha}\text{-}\bigwedge)$

$$\leftrightarrow \{\bigwedge x\neg\neg\bigvee y\bigwedge zA(x,y,z) \rightarrow \neg\neg\bigwedge x\bigvee y\bigwedge zA(x,y,z)\} \qquad \text{(P16)}$$

This by full $(\overset{\neg}{\bigwedge}\overset{\neg}{})^{o}\text{-}\bigvee\bigwedge$.

(3.12) (ωAC)-analysis $\vdash A \Leftrightarrow (\omega AC)$-Heyting-analysis plus (MP), $(\overset{\neg\neg}{\omega}AC)\text{-}\bigwedge \vdash \neg\neg A^{*}$

$$(\omega AC)\text{-analysis} \vdash A \Longleftrightarrow (\omega AC)\text{-Heyting-analysis plus (MP)},(\underset{\rightarrow}{V}),\ (\overset{\neg\neg}{\omega}AC)\text{-}\Lambda \vdash \neg\neg A^*$$

$$\Longrightarrow \text{Heyting-analysis plus (MP)},(\underset{\rightarrow}{V}),\ (\overset{\neg\neg}{\omega}AC)\text{-}\Lambda \vdash \neg\neg A^*$$

Proof:

Induction on the length of proof. (3.8) - (3.10). By chapter I (ωAC)-Λ suffices.

$$\neg\neg((\omega AC)^{\alpha}-\Lambda)^* \equiv \neg\neg\{\Lambda x^{o}\neg\neg\Lambda y^{\alpha}\neg\neg Vz^{\alpha}\Lambda z_1\neg\neg A(x,y,z,z_1) \longrightarrow Vx^{\alpha o}\Lambda y^{o}\neg\neg\Lambda z_1\neg\neg A(y,xy,x(y'),z_1)\} \quad \text{(A quantifierfree)}$$

$$\leftrightarrow\{\Lambda x,y\neg\neg Vz\Lambda z_1 A(x,y,z,z_1) \longrightarrow \neg\neg Vx\Lambda y,z_1 A(y,xy,x(y'),z_1)\} \quad (\neg\neg\Lambda\neg\neg \longleftrightarrow \Lambda\neg\neg, (P16))$$

$$\equiv (\overset{\neg\neg}{\omega}AC)^{\alpha}\text{-}\Lambda$$

(3.13) Let K_1, K_2 be formal systems with:

(1) $K_1 \vdash A \rightleftarrows K_2 \vdash \neg\neg A^*$

(2) $K_2 \subset K_1$

(3) All prime formulae of K_2 are decidable.

(4) K_2 contains intuitionistic predicate logic.

(a) $K_1 \vdash A \rightleftarrows K_2 \vdash A$ for A quantifierfree or $A \equiv \neg V\underline{\underline{x}}_1 \Lambda \underline{\underline{y}}_1 .. V\underline{\underline{x}}_n \Lambda \underline{\underline{y}}_n B$ (B quantifierfree) (Kreisel [24])

(b) Contains K_2 moreover (MP), functional contraction (1.1) (or (MP) for tupels), so (a) also holds for $A \equiv \Lambda\underline{\underline{x}}V\underline{\underline{y}}B$ (B quantifierfree).

Proof:

A quantifierfree: $K_1 \vdash A \rightleftarrows K_2 \vdash \neg\neg A \Longleftrightarrow K_2 \vdash A$ (1),(3),(4)

$A \equiv \neg V\underline{\underline{x}}_1 \Lambda \underline{\underline{y}}_1 ... V\underline{\underline{x}}_n \Lambda \underline{\underline{y}}_n B$:

$$\neg\neg A^* \equiv \neg\neg\neg V\underline{\underline{x}}_1 \Lambda y_{11}\neg\neg ... \Lambda y_{1m_1}\neg\neg ... V\underline{\underline{x}}_n \Lambda y_{n1}\neg\neg ... \Lambda y_{nm_n}\neg\neg B$$

(5) $\rightarrow \neg V\underline{\underline{x}}_1 \Lambda \underline{\underline{y}}_1 ... V\underline{\underline{x}}_n \Lambda \underline{\underline{y}}_n B \equiv A$ (4), $A \rightarrow \neg\neg A$

Thus: $K_1 \vdash A \rightleftarrows K_2 \vdash \neg\neg A^* \Longleftrightarrow K_2 \vdash A$ (1),(4),(5),(2)

$A \equiv \Lambda\underline{\underline{x}}V\underline{\underline{y}}B$:

(6) $\neg\neg A^* \equiv \neg\neg\Lambda x_1\neg\neg ... \Lambda x_m\neg\neg V\underline{\underline{y}}B \leftrightarrow \Lambda\underline{\underline{x}}\neg\Lambda\underline{\underline{y}}\neg B \leftrightarrow \Lambda\underline{\underline{x}}V\underline{\underline{y}}B \equiv A$ (4),(MP),(3)

Thus: $K_1 \vdash A \rightleftarrows K_2 \vdash \neg\neg A^* \Longleftrightarrow K_2 \vdash A$ (1),(6),(4)

Unlike number theory (3.8) for analysis (3.11), (3.12) something must be added to the direct intuitionistic pendant because in $(AC)^{0}\text{-}\bigwedge$, $(\omega AC)\text{-}\bigwedge$ the $\bigwedge$-quantifiers of the premises are (seen constructively) weakened by the $\neg\neg^{*}$-translation. Actually full Heyting-analysis plus (MP), $(\underset{\rightarrow}{\vee})$ is proof-theoretically not stronger than number theory (see chapter VI, X).

For analysis with comprehension only this difference vanishes formally.

(3.14) Intuitionistically plus $\neg\neg(C)^{\alpha}$: $\neg\neg((C)^{\alpha})^{*}$ (Kreisel [27],153)

<u>Proof:</u>

$$\neg\neg(C)^{*} \equiv \neg\neg\bigvee y^{o\alpha}\bigwedge x^{\alpha}\neg\neg(yx \overset{o}{=} 0 \leftrightarrow A^{*}(x)) \leftrightarrow \neg\neg\bigvee y\bigwedge x(yx=0 \leftrightarrow \neg\neg A^{*}(x)) \quad \text{(P16)}$$

This by $\neg\neg(C)^{\alpha}$.

Therefore with analogous notions as previously:

$$(C)\text{-analysis} \vdash A \Longleftrightarrow (C)\text{-Heyting-analysis} \vdash \neg\neg A^{*}$$
$$\Longleftrightarrow \neg\neg(C)\text{-Heyting-analysis} \vdash \neg\neg A^{*}$$

Actually comprehension has in Heyting-analysis the following position.

(3.15) Heyting-analysis $\vdash$ $(C) \leftrightarrow \bigwedge(\text{Tnd})$ (Kreisel [23],108)

<u>Proof:</u>

$$\bigwedge x^{\alpha}(A(x) \vee \neg A(x)) \leftrightarrow \bigwedge x^{\alpha}\bigvee y^{o}(y=0 \leftrightarrow A(x)) \quad \text{(P4)}$$
$$\leftrightarrow \bigvee z^{o\alpha}\bigwedge x^{\alpha}(zx=0 \leftrightarrow A(x)) \quad (AC)^{\alpha,o}$$

(3.16) Intuitionistically: $\neg\neg((AC)^{\alpha,o})^{*} \vdash \neg\neg((C)^{\alpha})^{*}$

<u>Proof:</u> Since $(AC)^{\alpha,o} \vdash (C)^{\alpha}$ classically, so (3.16) follows with (3.6).

(3.17) Heyting-analysis: $\neg\neg(C)^{\alpha} \vdash \neg\neg((AC)^{\alpha,\beta})^{*}$

<u>Proof:</u>

$$\neg\neg(C)^{\alpha} \leftrightarrow \neg\neg\bigwedge x^{\alpha}(A(x) \vee \neg A(x)) \quad (3.15)$$
$$\vdash \bigwedge x^{\alpha}\neg\neg A(x) \rightarrow \neg\neg\bigwedge x^{\alpha}A(x)$$
$$\vdash \bigwedge x^{\alpha}\neg\neg\bigvee y^{\beta}A^{*}(x,y) \rightarrow \neg\neg\bigvee z^{\beta\alpha}\bigwedge x^{\alpha}\neg\neg A^{*}(x,zx) \quad (AC)^{\alpha,\beta}$$

(3.18) Heyting-analysis plus (MP):

$\neg\neg((C)^{\gamma o})^* \vdash \neg\neg((AC)^{\alpha,\beta})^*$

$\neg\neg\bigwedge x^{\gamma o}(\bigvee y A(x,y) \vee \neg\bigvee y A(x,y)) \vdash \neg\neg((AC)^{\alpha,\beta}\text{-}\bigwedge)^*$ (A quantifierfree)

γ from α,β as in (1.1)

<u>Proof:</u>

(1) $\neg\neg((C)^{\gamma o})^* \longleftrightarrow \neg\neg\bigwedge x^{\gamma o}(\neg\neg A^*(x) \vee \neg\neg\neg A^*(x))$ proof (3.14),(3.15)

$\bigwedge x^{\alpha}\neg\neg\bigvee y^{\beta}A^*(x,y) \longrightarrow \neg\neg\bigwedge x\neg\bigwedge y\neg A^*(x,y)$

$\longrightarrow \neg\neg\bigwedge x\bigvee y\neg\neg A^*(x,y)$ $((MP)^{\gamma}$ for $\neg A^*$,(1),(1.1), premise)

$\longrightarrow \neg\neg\bigvee z^{\beta\alpha}\bigwedge x^{\alpha}\neg\neg A^*(x,zx)$ $(AC)^{\alpha,\beta}$

From this $\neg\neg((AC)^{\alpha,\beta})^*$. - $\neg\neg((AC)^{\alpha,\beta}\text{-}\bigwedge)^*$ follows by the preceding from $\neg\neg\bigwedge x^{\alpha},y^{\beta}(\neg\bigwedge z\neg\neg A(x,y,z) \vee \neg\neg\bigwedge z\neg\neg A(x,y,z))$, A quantifierfree. This is obtained from the deduction premise with (1.1).

After (3.11), (3.12) thus by means of $\neg\neg^*$ we have an interpretation of $(AC)^o$- resp. (ωAC)-analysis in $(AC)^o$-Heyting-analysis plus (MP), $(\underset{\rightarrow}{\bigvee})$, $(\neg\bigwedge\neg)^o$-$\bigvee\bigwedge$ resp. (ωAC)-Heyting-analysis plus (MP), $(\underset{\rightarrow}{\bigvee})$, $(\overset{\neg\neg}{\omega}AC)$-$\bigwedge$. The restriction to $\bigvee\bigwedge$ resp. $\bigwedge$ is inessential for the corresponding general cases (see corollary to (1.1), (1.2)).

IV. Gödel's functional interpretation in the narrower sense

The functional interpretation in the narrower sense associates with each formula of the intuitionistic approximated theories of chapter III a $\bigvee\bigwedge$-formula. Later on for provable formulae these $\bigvee$-magnitudes are effectively exhibited in a functional calculus. The functional interpretation in the general sense consists of all these processes: intuitionistic $\neg\neg^*$-approximation, functional interpretation in the narrower sense and functional exhibition. From this then a consistency proof arises by computing all closed type 0 terms of the functional calculus in a standard way so that all closed equations which are provable in the calculus when reduced to type 0 pass into 'valid' equations thus excluding the contradiction 0=1. (In presence of higher intensional equalities $\overset{\alpha}{=}$ one has to proceed in a similar way with respect to a reduction of all terms (see Tait [37]).)

Here we refer without exception to the functional language. Other languages can be handled similarly when functionals are added.

qct(A) denotes the contraction of <u>all</u> rows of homogeneous quantifiers from within A by (1.1).

Inductive definition of the functional interpretation ':

I. $A' \equiv: A$ for quantifierfree A

II. For formulae not coming under I with $A' \equiv \bigvee x_1 \bigwedge x_2 A_o(x_1,x_2)$, $B' \equiv \bigvee y_1 \bigwedge y_2 B_o(y_1,y_2)$ (A_o, B_o quantifierfree) the following is stipulated:

$(A \wedge B)' \equiv: \text{qct}\{\bigvee x_1,y_1 \bigwedge x_2,y_2(A_o(x_1,x_2) \wedge B_o(y_1,y_2))\}$

$(A \vee B)' \equiv: \text{qct}\{\bigvee z^o,x_1,y_1 \bigwedge x_2,y_2((z=0 \wedge A_o(x_1,x_2)) \vee (z\neq 0 \wedge B_o(y_1,y_2)))\}$

$(\bigvee xA(x))' \equiv: \text{qct}\{\bigvee x,x_1 \bigwedge x_2 A_o(x_1,x_2,x)\}$

$(\bigwedge xA(x))' \equiv: \text{qct}\{\bigvee z_1 \bigwedge x,x_2 A_o(z_1x,x_2,x)\}$

$(A \rightarrow B)' \equiv: \text{qct}\{\bigvee z_1,z_2 \bigwedge x_1,y_2(A_o(x_1,z_2x_1y_2) \rightarrow B_o(z_1x_1,y_2))\}$

$(\neg A)' \equiv: \bigvee z_2 \bigwedge x_1 \neg A_o(x_1,z_2x_1)$

In case of lacking quantifiers one has to cross out all magnitudes which refer to these arguments.

This definition differs from the original version of Gödel by the treatment of quantifierfree formulae and by using qct. As already mentioned in chapter I thereby we abbreviate correctly the intensional tupel-notation; because for analysis this notation is in shortened form too opaque and in full detail too extensive. It will turn out that the needed (ER)-qf gives no proof-theoretic strengthening; moreover extensionality for sequences is in any case necessary for analysis. - As later on particular functionals must be persuited throughout the '-formation a total quantifier contraction is not at all places practical because the coding ψ_o, ψ_1, ϕ must always be taken into consideration. Here it is recommendable to use a combination of tupel formation and functional contraction depending on the individual case. The admissibility is shown in (4.1). For that purpose we need the notions "$\underline{\varphi}$-solution" and "variant" of a formula.

$\underline{\varphi}\text{-sol}(\bigvee \underline{x} \bigwedge \underline{y} A(\underline{x},\underline{y},\underline{u})) \equiv: A(\underline{\underline{\varphi u}},\underline{v},\underline{u})$ where A quantifierfree, $\underline{u}$ the stock of free variables in $\bigvee \underline{x} \bigwedge \underline{y} A(\underline{x},\underline{y},\underline{u})$, $\underline{v}$ new free variables

var(A') $\equiv$: '-formation of A with arbitrary (possibly zero) quantifier contractions; now in the above definition the bound variables must be read as tupels.

(4.1) Over the calculus of the primitive recursive functionals (i.e. the common quantifierfree part of the formal systems of analysis plus (ER)-qf and term substitution) for each φ-sol(A') there exists effectively a provably equivalent $\underline{\psi}$-sol(var(A')) and vice versa.

<u>Proof:</u> We compare the processes of A'- and var(A')-formation from the inside in all logical details (prenexing, choice). At places which differ by functional contraction (1.1) the corresponding primitive recursive correlations between the magnitudes involved are set up. These correlations are hereditary step by step by including the new magnitudes. As an example we take choice and negation applications:

$\bigwedge x_1,x_2 \bigvee y_1,y_2 A(x_1,x_2,y_1,y_2)$ $\qquad$ $\bigwedge x \bigvee y A(\psi_o x,\psi_1 x,\psi_o y,\psi_1 y)$

$\bigvee z_1,z_2 \bigwedge x_1,x_2 A(x_1,x_2,z_1x_1x_2,z_2x_1x_2)$ $\qquad$ $\bigvee z \bigwedge x A(\psi_o x,\psi_1 x,\psi_o(zx),\psi_1(zx))$

Here we have the following correspondences:

$\Rightarrow$: $x_1:\psi_o x$; $x_2:\psi_1 x$; $zx:\phi(z_1(\psi_o x)(\psi_1 x),z_2(\psi_o x)(\psi_1 x))$

$\Leftarrow$: $x:\phi(x_1,x_2)$; $z_1x_1x_2:\psi_o(z(\phi(x_1,x_2)))$; $z_2x_1x_2:\psi_1(z(\phi(x_1,x_2)))$

For negationing only the directions of the correspondences change.

(4.2)

(a) Heyting-analysis $\vdash (A\wedge B)' \leftrightarrow A'\wedge B'$, $(A\vee B)' \leftrightarrow A'\vee B'$,

$(\bigvee xA(x))' \leftrightarrow \bigvee xA'(x)$, $(\bigwedge xA(x))' \leftrightarrow \bigwedge xA'(x)$,

$(A\rightarrow B)' \rightarrow (A'\rightarrow B')$, $(\neg A)' \rightarrow \neg A'$

(b) Heyting-analysis plus (MP) $\vdash (\neg A)' \leftrightarrow \neg A'$

(c) Heyting-analysis plus (MP), $(\xrightarrow{\bigvee})$ $\vdash (A\rightarrow B)' \leftrightarrow (A'\rightarrow B')$

(d) Over Heyting-arithmetic plus (ER)-qf:

(AC), (MP), $(\xrightarrow{\bigvee}) \dashv\vdash A \leftrightarrow A'$ (Yasugi [64])

(ER)-qf can be avoided by using intensional tupels.

Proof:

Ad (a):

$$(A\wedge B)' \leftrightarrow \bigvee x_1,y_1 \bigwedge x_2,y_2 (A_o(x_1,x_2) \wedge B_o(y_1,y_2)) \quad (4.1)$$
$$\leftrightarrow \bigvee x_1 \bigwedge x_2 A_o(x_1,x_2) \wedge \bigvee y_1 \bigwedge y_2 B_o(y_1,y_2) \equiv A'\wedge B'$$

$$(A\vee B)' \leftrightarrow \bigvee z^o,x_1,y_1 \bigwedge x_2,y_2 \{(z=0\wedge A_o(x_1,x_2)) \vee (z\neq 0\wedge B_o(y_1,y_2))\} \quad (4.1)$$
$$\leftrightarrow \bigvee z^o,x_1,y_1 \bigwedge x_2,y_2 \{(z=0\rightarrow A_o(x_1,x_2)) \wedge (z\neq 0\rightarrow B_o(y_1,y_2))\} \quad (P4)$$
$$\leftrightarrow \bigvee z^o,x_1,y_1 \{(z=0\rightarrow \bigwedge x_2 A_o(x_1,x_2)) \wedge (z\neq 0\rightarrow \bigwedge y_2 B_o(y_1,y_2))\}$$
$$\leftrightarrow \bigvee z^o,x_1,y_1 \{(z=0\wedge \bigwedge x_2 A_o(x_1,x_2)) \vee (z\neq 0\wedge \bigwedge y_2 B_o(y_1,y_2))\} \quad (P4)$$
$$\leftrightarrow \bigvee z^o \{(z=0\wedge \bigvee x_1 \bigwedge x_2 A_o(x_1,x_2)) \vee (z\neq 0\wedge \bigvee y_1 \bigwedge y_2 B_o(y_1,y_2))\}$$
$$\leftrightarrow \bigvee x_1 \bigwedge x_2 A_o(x_1,x_2) \vee \bigvee y_1 \bigwedge y_2 B_o(y_1,y_2) \equiv A'\vee B'$$

$$(\bigvee xA(x))' \leftrightarrow \bigvee x,x_1 \bigwedge x_2 A_o(x_1,x_2,x) \equiv \bigvee xA'(x) \quad (4.1)$$

$$(\bigwedge xA(x))' \leftrightarrow \bigvee z_1 \bigwedge x,x_2 A_o(z_1x,x_2,x) \quad (4.1)$$
$$\leftrightarrow \bigwedge x \bigvee x_1 \bigwedge x_2 A_o(x_1,x_2,x) \equiv \bigwedge xA'(x) \quad (AC)$$

$$(A\rightarrow B)' \leftrightarrow \bigvee z_1,z_2 \bigwedge x_1,y_2 \{A_o(x_1,z_2x_1y_2) \rightarrow B_o(z_1x_1,y_2)\} \quad (4.1)$$
$$\leftrightarrow \bigwedge x_1 \bigvee y_1 \bigwedge y_2 \bigvee x_2 \{A_o(x_1,x_2) \rightarrow B_o(y_1,y_2)\} \quad (AC)$$
$$\rightarrow \{\bigvee x_1 \bigwedge x_2 A_o(x_1,x_2) \rightarrow \bigvee y_1 \bigwedge y_2 B_o(y_1,y_2)\} \equiv (A'\rightarrow B')$$

$$(\neg A)' \leftrightarrow \bigvee z_2 \bigwedge x_1 \neg A_o(x_1,z_2x_1)$$

$$\begin{aligned}(\neg A)' &\leftrightarrow \bigwedge x_1 \bigvee x_2 \neg A_o(x_1,x_2) && \text{(AC)}\\ &\rightarrow \bigwedge x_1 \neg \bigwedge x_2 A_o(x_1,x_2) \\ &\leftrightarrow \neg \bigvee x_1 \bigwedge x_2 A_o(x_1,x_2) \equiv \neg A'\end{aligned}$$

<u>Ad (b):</u>
By (MP), (P16) in the last proof of (a) the arrow can be inverted.

<u>Ad (c):</u>

$$\begin{aligned}(A' \rightarrow B') &\equiv (\bigvee x_1 \bigwedge x_2 A_o(x_1,x_2) \rightarrow \bigvee y_1 \bigwedge y_2 B_o(y_1,y_2)) \\ &\leftrightarrow \bigwedge x_1 \bigvee y_1 \bigwedge y_2 \neg \bigwedge x_2 (A_o(x_1,x_2) \wedge \neg B_o(y_1,y_2)) && (\underset{\rightarrow}{V}),\text{(P16)} \\ &\leftrightarrow \bigwedge x_1 \bigvee y_1 \bigwedge y_2 \bigvee x_2 \{A_o(x_1,x_2) \rightarrow B_o(y_1,y_2)\} && \text{(MP),(P16)} \\ &\leftrightarrow (A \rightarrow B)' && \text{proof (a)}\end{aligned}$$

<u>Ad (d):</u>
(AC),(MP),$(\underset{\rightarrow}{V}) \vdash A \leftrightarrow A'$ follows from (a) - (c) by induction on the F-rank of A. The converse ensues from the primitive recursive functional solution of (AC)', (MP)', $(\underset{\rightarrow}{V})'$ given in chapter VI.

It is shown in chapter VI that the functional interpretation in the narrower sense of Heyting-analysis plus (MP), $(\underset{\rightarrow}{V})$ can effectively be given in the calculus of the primitive recursive functionals. - So already over this functional domain functional interpretation in the narrower sense is not adequate for intuitionistic predicate logic (Kreisel [23],113; [25]. Kleene-Vesley [22],131). The corresponding question for intuitionistic propositional logic is still open.

Together with (4.2) we have for the functional interpretation of a classical theory K_1 with $\neg\neg^*$-equivalent K_2 in a functional calculus F containing the primitive recursive functionals:

$$\left.\begin{array}{l} K_1 \cup F \vdash A \rightleftarrows K_2 \cup F \vdash \neg\neg A^* \\ K_1 \vdash A \rightleftarrows K_2 \vdash \neg\neg A^* \end{array}\right\} \Longrightarrow K_2 \cup F \cup \text{Heyting-analysis} \cup (MP) \cup (\underset{\rightarrow}{V}) \vdash \neg\neg A^*$$

$$\Longleftrightarrow \bigvee \varphi \varepsilon F \;_F\vdash \varphi\text{-sol}(\neg\neg A^*)' \qquad (4.2)$$

φ-sol$(\neg\neg A^*)'$ reflects in this way the constructive content of A. - For our formal systems of analysis the functional interpretation (in the narrower sense) of Heyting-analysis plus (MP), $(\underset{\rightarrow}{V})$ requires in chapter VI no other means than those already necessary for the corresponding K_2 of chapter III. One may therefore conjecture (analogous to arithmetic, see chapter VI) that in our cases all formal systems standing above uppermost have the same proof-theoretic strength.

(4.3) The functional interpretation embraces the no-counterexample-interpretation, which associates to each classically provable prenex formula $\bigwedge x_1 \bigvee y_1 .. \bigwedge x_n \bigvee y_n A(x_1,..,x_n,y_1,..,y_n)$ (A quantifierfree) a constructive solution of $\bigvee z_1,..,z_n \bigwedge x_1,x_2,..,x_n A(x_1,x_2(z_1x_1..x_n),.., x_n(z_1x_1..x_n)..(z_{n-1}x_1..x_n),z_1x_1..x_n,..,z_nx_1..x_n)$. (Kreisel [23],113)

Proof:

For classically provable $\bigwedge x_1 \bigvee y_1 .. \bigwedge x_n \bigvee y_n A(x_1,..,x_n,y_1,..,y_n)$
$\neg\neg\bigwedge x_1 \neg\neg\bigvee y_1 .. \bigwedge x_n \neg\neg\bigvee y_n A(x_1,..,x_n,y_1,..,y_n)$ is derivable in the intuitionistic $\neg\neg^*$-approximation . Because of the intuitionistic relations

$$\bigvee x_1 \bigwedge y_1 .. \bigvee x_n \bigwedge y_n \neg A \rightarrow \overline{\neg\bigwedge x_1 \underline{\neg\bigwedge y_1} \neg} .. \overline{\neg\bigwedge x_n \underline{\neg\bigwedge y_n} \neg} A$$
$$\leftrightarrow \neg\bigwedge x_1 \neg\neg\bigvee y_1 .... \bigwedge x_n \neg\neg\bigvee y_n A$$

the latter contains also $\neg\bigvee x_1 \bigwedge y_1 .. \bigvee x_n \bigwedge y_n \neg A(x_1,..,x_n,y_1,..,y_n)$.
The functional interpretation ' for this is

$$(\neg\bigvee x_1,..,x_n \bigwedge y_1,..,y_n \neg A(x_1,x_2y_1,..,x_ny_1...y_{n-1},y_1,..,y_n))' \equiv$$
$$\bigvee z_1,..,z_n \bigwedge x_1,..,x_n \cancel{\neg\neg} A(x_1,x_2(z_1x_1..x_n),..,x_n(z_1x_1..x_n)..(z_{n-1}x_1..x_n), z_1x_1..x_n,..,z_nx_1..x_n)$$

Because in classical theories over the functional language each formula is provably equivalent to a $\bigwedge\bigvee...\bigwedge\bigvee$-formula by prenexation and trivial quantifier additions, so the no-counterexample part of functional interpretation already reflects in this way the constructive content. However for our formal systems of analysis a more direct correlation can be given. To that purpose we consider those formulae which after (3.13), (3.11), (3.12) are invariant with respect to $\neg\neg^*$.

Quantifierfree formulae are not changed by the '-functional interpretation. Therefore:

(4.4) The functional calculus of the interpretation covers the quantifierfree functional part of the original classical theory.

For formulae $A \equiv \neg\bigvee x_1 \bigwedge y_1 \ldots \bigvee x_n \bigwedge y_n B$ (B quantifierfree) it follows by the proof of (4.3) that sol(A') is the no-counterexample-interpretation of $\bigwedge x_1 \bigvee y_1 \ldots \bigwedge x_n \bigvee y_n \neg B$.

If $A \equiv \bigwedge x \bigvee y B$ (B quantifierfree) we have $A' \equiv \bigvee z \bigwedge x B(x,zx)$. For provable A these z are effectively exhibited by functional interpretation. Thus:

(4.5) If the functional interpretation can be carried out, only such functionals can be proved in $\bigwedge\bigvee$-form to exist in the original classical theory which are also in the corresponding functional calculus of the interpretation. - Because in our formal systems of analysis by (1.2) each formula is provably equivalent to a $\bigwedge\bigvee$-formula, in this way the constructive content of provable formulae is directly expressed by effective functional operations.

At first one could tend to the opinion that at least for pure logical statements this functional interpretation depends only on the connectives but not on the prime formulae. This is not the case.

(4.6) Gödel's functional interpretation in the narrower sense depends already for intuitionistic predicate logic on the prime formulae. Thus besides (4.4) the functional calculus of the interpretation has to allow the formation of characteristic functionals for quantifierfree formulae (of the original classical theory), i.e. all employed prime formulae must be decidable.

Proof:

Consider the following '-functional interpretation:

$$(\neg\neg(\bigvee x A(x) \vee \neg\bigvee y A(y)))' \equiv (\neg\neg\bigvee z^0, x \bigwedge y\{(z=0 \wedge A(x)) \vee (z \neq 0 \wedge \neg A(y))\})'$$

(A quantifierfree)
$$\equiv (\neg\bigvee y_1 \bigwedge z^0, x \neg\{(z=0 \wedge A(x)) \vee (z \neq 0 \wedge \neg A(y_1 zx))\})'$$
$$\equiv \bigvee z_1, x_1 \bigwedge y_1 \neg\neg\{(z_1 y_1 = 0 \wedge A(x_1 y_1)) \vee (z_1 y_1 \neq 0 \wedge \neg A(y_1(z_1 y_1)(x_1 y_1)))\}$$

If $A(v) \leftrightarrow \curlyvee$, so $z_1 y_1 \equiv 0$; if $A(v) \leftrightarrow \curlywedge$, so $sg(z_1 y_1) \equiv 1$. Thus the z_1-solution depends on A. - A primitive recursive fulfillment for instance is:

$$z_1 y_1 \stackrel{\circ}{=}: \begin{cases} 1 & \text{if } \neg A(y_1 1 \sigma) \\ 0 & \text{otherwise} \end{cases} \qquad x_1 y_1 =: \begin{cases} \sigma & \text{if } \neg A(y_1 1 \sigma) \\ y_1 1 \sigma & \text{otherwise} \end{cases}$$

To carry this out in the calculus of the primitive recursive functionals at first one has to form the characteristic functional φ_A of A: $\varphi_A v=0 \leftrightarrow A(v)$.

Remark:
Diller and Nahm [6] give a modification ^ of Gödel's '-functional interpretation which at least for number theory needs no characteristic functional:

$$(\bigvee v \bigwedge w A_o(v,w) \rightarrow \bigvee y \bigwedge z B_o(y,z))^\wedge \;\equiv:\; \bigvee Y,X,W \bigwedge v,z(\bigwedge x^o{<}XvzA_o(v,Wvzx) \rightarrow B_o(Yv,z)) \quad (A_o,B_o \text{ quantifierfree})$$

$\neg$ accordingly; the other connectives are treated as in Gödel's '-interpretation.

Here we don't enter into this ^-interpretation because we have only decidable prime formulae, and we therefore get through with the shorter '-interpretation. However for theories with undecidable prime formulae (e.g. higher intensional equalities) a constructive functional interpretation can only be given by such ^-interpretations not depending on the prime formulae.

(4.7) An enumerable formal system K over the functional language which is functional interpretable in a (quantifierfree) functional calculus F with the properties (a) - (e) is ω-consistent.

(a) Each closed type 0 term of F has a standard computation.

(b) Let t_i^α be the ith term in an effective denumeration of a set of F-terms of type α with a certain stock $v_1,\ldots,v_m$ of variables. - Each extension F^+ of F by functionals ϕ with the computation instructions (+) $\phi v_1 \ldots v_m z^o \overset{\alpha}{=}: t_{f(z)}$ for all numerals z and a computable 00-function f - has for type 0 terms of F^+ a standard computation, namely the extension of the F-computation (a) by (+).

(c) F contains the zero-functionals, term substitution and the equality relations.

(d) The rules of F have the property that for all applications over F^+ the F^+-computation (b) of a closed conclusion gives a "valid" statement provided the same holds for suitable F^+-closed instances of the corresponding premises.

(e) F^+-closed instances of the axioms of F pass via F^+-computation (b) into "valid" statements.

For example the calculus of the primitive recursive functionals has these properties (a) - (e).

<u>Proof:</u>

(1) In F, F^{+} only "valid" formulae are provable. This follows by induction on the length of proof. For F-axioms this holds by (e), (b) and for the additional axioms (+) by (b). The rules transmit this property by (d), (c), (b).

(2) Therefore F, F^{+} are consistent.

Under the assumptions (3) K is effectively functional interpretable in F and (4) K is enumerable we have to show that not at the same time (5) ${}_{K}\vdash A(z)$ for all numerals z and (6) ${}_{K}\vdash \neg\bigwedge x^{o}A(x)$.

We assume (5), (6) and produce a contradiction. - K_i be the enumerable intuitionistic $\neg\neg^{*}$-approximation of K according to chapter III.

(7) ${}_{K_i}\vdash \neg\neg A^{*}(z)$ for all numerals z (3),(5)

(8) ${}_{K_i}\vdash \neg\neg\neg\bigwedge x^{o}\neg\neg A^{*}(x)$ (3),(6)

Take $(\neg\neg A^{*}(z^{o}))' \equiv \bigvee x_1^{\alpha}\bigwedge y_1^{\beta}A_o(x_1,y_1,z,\underline{\underline{u}})$ (A_o quantifierfree, type/$\underline{\underline{u}}\equiv\underline{\underline{y}}$)

$(\bigwedge x^{o}\neg\neg A^{*}(x))' \equiv \bigvee x_2^{\alpha o}\bigwedge x^{o},y_1^{\beta}A_o(x_2x,y_1,x,\underline{\underline{u}})$

$(\neg\bigwedge x^{o}\neg\neg A^{*}(x))' \equiv \bigvee x_3^{o(\alpha o)},y_2^{\beta(\alpha o)}\bigwedge x_2^{\alpha o}\neg A_o(x_2(x_3x_2),y_2x_2,x_3x_2,\underline{\underline{u}})$

By (7), (3), (4), (b) to each numeral z there is effectively a closed term $\varphi_{f(z)}^{\alpha\underline{\underline{y}}}\varepsilon F$ (f^{oo} computable function) with

(9) ${}_{F}\vdash A_o(\varphi_{f(z)}\underline{\underline{u}}, v^{\beta}, z, \underline{\underline{u}})$.

Moreover by (8), (3) there are functionals $\psi_1^{o(\alpha o)\underline{\underline{y}}}$, $\psi_2^{\beta(\alpha o)\underline{\underline{y}}}$ ε F with

(10) ${}_{F}\vdash \neg A_o(w^{\alpha o}(\psi_1\underline{\underline{u}}w), \psi_2\underline{\underline{u}}w, \psi_1\underline{\underline{u}}w, \underline{\underline{u}})$.

Let F^{+} be the following (b)-extension of F by the functional ϕ:

(11) ${}_{F^{+}}\vdash \phi^{\alpha o\underline{\underline{y}}}\underline{\underline{u}}z^{o} =: \varphi_{f(z)}\underline{\underline{u}}$ for all numerals z

Term substitution $w:\phi\underline{\underline{\sigma}}$, $\underline{\underline{u}}:\underline{\underline{\sigma}}$ in (10) yields

(12) ${}_{F^{+}}\vdash \neg A_o(\phi\underline{\underline{\sigma}}(\psi_1\underline{\underline{\sigma}}(\phi\underline{\underline{\sigma}})), \psi_2\underline{\underline{\sigma}}(\phi\underline{\underline{\sigma}}), \psi_1\underline{\underline{\sigma}}(\phi\underline{\underline{\sigma}}), \underline{\underline{\sigma}})$ (11),(c),(b)

Let $\bar{z}$ be the computation result of $\psi_1\underline{\underline{\sigma}}(\phi\underline{\underline{\sigma}})$ according to (b):

(13) $\bar{z} \overset{o}{=} \psi_1\underline{\underline{\sigma}}(\phi\underline{\underline{\sigma}})$

(14) $F^+ \vdash \neg A_0(\varphi_{f(\bar{z})}\underline{\sigma}, \psi_2\underline{\sigma}(\phi\underline{\sigma}), \bar{z}, \underline{\sigma})$ (12),(13),(11),(c)

But from (9) follows by term substitution

(15) $F^+ \vdash A_0(\varphi_{f(\bar{z})}\underline{\sigma}, \psi_2\underline{\sigma}(\phi\underline{\sigma}), \bar{z}, \underline{\sigma})$ (c),(11)

With (14) this contradicts (2).

Remark:

1. The extension of the (4.7)-argumentation to higher types - in order to show that $_{K,K_i} \vdash A(t_j^\alpha)$ for all closed terms $t_j^\alpha \in K,K_i$ and $_{K,K_i} \vdash \neg \bigwedge x^\alpha A(x)$ are not possible at the same time - needs the additional assumptions: $F = F^+$ = quantifierfree part of K,K_i. Also the ϕ-functionals (+) are now without further arguments not extensional or continuous.

2. To functional calculi F, F^+ with the above properties (a) - (e) one can add consistently the quantifierfree ω-rule and all generalizations $(\omega R)^\alpha$ to higher types

$$(\omega R)^\alpha \quad \frac{A(t_1^\alpha),\ A(t_2^\alpha),\dots \quad \text{for all closed terms } t_j^\alpha \text{ of the calculus}}{A(v^\alpha)}$$

The argumentation for (1), (2) covers also these rules because a closed conclusion reduces immediately to one of the premises.

The functional interpretation in the general resp. narrower sense of a constructive $(\omega R)^\alpha$-rule in K resp. K_i can be given in the case $\alpha \equiv 0$ by (b), (c), (9), (11), $(\omega R)^0$ in constructive $^+$- and $(\omega R)^0$-iterations starting with F (see Schütte [46],168-171). For higher types α we need - as in 1. - also here the addition: $F = F^+$ = quantifierfree part of K,K_i.

V. The calculus T of the primitive recursive functionals

The calculus of the primitive recursive functionals was previously introduced as the common quantifierfree part of our formal systems of analysis plus (ER)-qf and term substitution. Now we will give a more detailed and for the further considerations useful development.

The language of T is the functional language of chapter I without quantifiers. $r \overset{\alpha}{=} s$, $\alpha \equiv 0\alpha_n \ldots \alpha_1$ denotes here $ru_1^{\alpha_1} \ldots u_n^{\alpha_n} \overset{0}{=} su_1 \ldots u_n$ where <u>the free variables $u_1, \ldots, u_n$ must be taken entirely new for each particular $\overset{\alpha}{=}$-sign.</u> - T has the following deduction frame.

Axioms of T

(T0) Relevant laws of intuitionistic propositional logic. $u \overset{\alpha}{=} u$.

(T1) $\varphi u_1 \ldots u_n \overset{\alpha}{=} u_1$ $\quad (n \geq 1)$

(T2) $\varphi u_1 \ldots u_n \overset{\alpha}{=} \psi$ $\quad$ for preceding ψ (including 0,')

(T3) $\varphi u_1 \ldots u_n \overset{\alpha}{=} \psi u_1 \ldots u_n (\chi u_1 \ldots u_n)$ $\quad$ for preceding ψ, χ $(n \geq 0)$

(T4) $\begin{cases} \varphi\, 0 \overset{\alpha}{=} \psi \\ \varphi u' = \chi(\varphi u)u^0 \end{cases}$ $\quad$ for preceding ψ, χ

Rules of T

(Mp) $\dfrac{A,\ A \rightarrow B}{B}$ $\qquad$ (S) $\dfrac{A(u^\alpha)}{A(t^\alpha)}$ if $u \notin A(0)$ (term substitution)

(E) $\dfrac{A \rightarrow r \overset{\alpha}{=} s}{A \wedge B(r) \rightarrow B(s)}$ $\qquad$ (extensionality)

(CI) $\dfrac{A(0),\ A(u^0) \rightarrow A(u')}{A(u)}$ if $u \notin A(0)$

An (E)-application within a proof is called <u>intensional</u> iff the proof of the premise $A \rightarrow r \overset{\alpha}{=} s$ uses only intensional (E)-applications and no special reference is made to the arguments behind the denoted terms r,s; for short, if $\overset{\alpha}{=}$ can be considered as an undefined notion having only the usual (intensional) equality properties and no further use is made of the above introduced $\overset{\alpha}{=}$-abbreviation. - The following derivations are foremost all intensional in this sense.

(T5) $\dfrac{A \rightarrow r \overset{\alpha}{=} s}{A \rightarrow t(r) \overset{0}{=} t(s)}$, $u \overset{0}{=} v \rightarrow v = u$, $u \overset{0}{=} v \wedge v = w \rightarrow u = w$; from this

$u \overset{0}{=} v \wedge A(u) \rightarrow A(v)$, $u \overset{0}{=} v \rightarrow r(u) \overset{0}{=} r(v)$.

Proof: (TO), (Mp) with (E) for $B(u^{\alpha}) \equiv : (t(r) \overset{0}{=} t(u))$ resp. $A \equiv : (u \overset{0}{=} v)$, $r \equiv : u$, $s \equiv : v$, $B(v) \equiv : (v \overset{0}{=} u)$ and $B(u) \equiv : (u \overset{0}{=} w)$.

Conversely (E) follows by induction on the F-rank of B from (T5). Thus

(5.1) (E) $\equiv$ (ER)-qf is in T deductive equivalent to (T5).

(5.2) Let $t_1(u_1, \ldots, u_m)^{\alpha}$ be a term containing only preceding functionals and at most the variables $u_1, \ldots, u_m$. There is a functional φ in T with: $\varphi u_1 \ldots u_m \overset{\alpha}{=} t(u_1, \ldots, u_m)$; T and all extensions are closed against (R1).

Proof: Induction on the T-rank of t

I. $t(u_1, \ldots, u_m)^{\alpha} \equiv u_i$ or ψ $(1 \leq i \leq m$, ψ a preceding functional)

1. $\varphi_1 u_i \ldots u_m \overset{\alpha}{=} u_i$ (T1)

$\varphi u_1 \ldots u_{i-1} = \varphi_1$ (T2)

Thus: $\varphi u_1 \ldots u_m \overset{\alpha}{=} \varphi_1 u_i \ldots u_m \overset{\alpha}{=} u_i$ (T5),(S),(TO),(Mp)

2. $\varphi u_1 \ldots u_m \overset{\alpha}{=} \psi$ (T2)

II. $t(u_1, \ldots, u_m)^{\alpha} \equiv r(u_1, \ldots, u_m)^{\beta}(s(u_1, \ldots, u_m)^{\gamma})$

By induction hypothesis there are functionals φ_1, φ_2 in T with:

$\varphi_1 u_1 \ldots u_m \overset{\beta}{=} r(u_1, \ldots, u_m)$, $\varphi_2 u_1 \ldots u_m \overset{\gamma}{=} s(u_1, \ldots, u_m)$.

Thus: $\varphi u_1 \ldots u_m \overset{\alpha}{=} \varphi_1 u_1 \ldots u_m(\varphi_2 u_1 \ldots u_m)$ (T3)

$\overset{\alpha}{=} r(u_1, \ldots, u_m)(s(u_1, \ldots, u_m)) \equiv t(u_1, \ldots, u_m)$ (E),(T5),(TO),(Mp)

Now in the usual notation the quantifierfree recursive functional theory is developed to the needed extent. - At first a list of T-results from chapter I.

(δ) $\delta 0 \overset{0}{=} 0$, $\delta u' = u$

(T6) $u' \overset{0}{\neq} 0$ (P1)

(T7) $u' = v' \leftrightarrow u \overset{0}{=} v$ (P2),(T5)

(T8) $u \neq 0 \leftrightarrow u = (\delta u)'$ (P3),(T6)

(T9) $u \overset{o}{=} 0 \vee u \neq 0$ (P4)

(I) $Iu \overset{o}{=} u$

(ν) $\nu uv \overset{o}{=} (uv)'$

(+) $+0 \overset{oo}{=} I, \quad +u' = \nu(+u^{o})$

(T10) $+uv' = +u'v = (+uv)'$ (P5)

(T11) $+uv = +vu$ (P6)

(T12) $u=0 \wedge v=0 \leftrightarrow +uv=0$ (P7)

(ε) $\varepsilon uv \overset{o}{=} \delta(uv)$

(-) $-0 \overset{oo}{=} I, \quad -u' = \varepsilon(-u^{o})$

<u>Definitions:</u> $r^{o} \leqslant s^{o} \;\equiv:\; s \geqslant r \;\equiv:\; (-sr \overset{o}{=} 0)$

$s < r \;\equiv:\; r > s \;\equiv:\; (-sr \overset{o}{\neq} 0)$

(T13) $u \geqslant v \vee u < v$ (P8)

(T14) $-u'v' = -uv, \quad u' \geqslant v' \leftrightarrow u \geqslant v, \quad u' > v' \leftrightarrow u > v$ (P9)

(T15) $u > v \rightarrow u = +v(-vu)$ (P10)

(T16) $-u(+u''''v) \neq 0$. With (T11): $u < u''''$ (P11)

(T17) $v \geqslant u \rightarrow v = +u(-uv)$ (P12)

(T18) The rule $\dfrac{A(0,v^{o}),A(u^{o},0),A(u,v) \rightarrow A(u',v')}{A(u,v)}$ $(u,v \notin A(0,0))$

is explicit derivable in T. (P13)

(T19) $u \overset{o}{=} v \vee u \neq v$ (P14)

(T20) $A \vee \neg A$ for quantifierfree A (P15)

(T21) All laws of classical propositional logic are provable in T.

(T22) $u \overset{o}{=} v \longleftrightarrow +(-uv)(-vu)=0$

<u>Proof:</u>
$A(u,v) \;\equiv:\; (u \overset{o}{=} v \leftrightarrow +(-uv)(-vu)=0)$
$u=0 \rightarrow +(-u0)(-0u)=+(-00)(-00)=0$
$+(-u0)(-0u)=0 \longrightarrow -0u=u=0$ (T12)
Thus A(u,0). Analogous A(0,v).
$A(u',v') \equiv (u'=v' \leftrightarrow +(-u'v')(-v'u')=0) \longleftrightarrow (u=v \leftrightarrow +(-uv)(-vu)=0) \equiv A(u,v)$
(T5),(T7),(T14)
(T22) follows now with (T18).

($\overline{\overline{sg}}$) $\overline{\overline{sg}}0 \overset{o}{=} 1, \quad \overline{\overline{sg}}u' = 0$ (5.2),(T4)

(T23) $u \neq 0 \leftrightarrow \overline{\overline{sg}}u = 0$

Proof:

$u \neq 0 \rightarrow u = (\delta u)'$ (T8)

$\quad \rightarrow \overline{\overline{sg}}u = \overline{\overline{sg}}(\delta u)' = 0$

$\overline{\overline{sg}}u = 0 \wedge u = 0 \rightarrow 1 = \overline{\overline{sg}}0 = \overline{\overline{sg}}u = 0$

(T24) For each quantifierfree $A(\underline{\underline{u}})$ ($\underline{\underline{u}}$ stock of variables in A) there exists in T a characteristic functional φ_A: $\varphi_A \underline{\underline{u}} \overset{o}{=} 0 \leftrightarrow A(\underline{\underline{u}})$.

Proof: Let $\widetilde{A}$ come from A by the usual classical reduction of $\vee, \rightarrow$ to $\neg, \wedge$. By (T21): $\widetilde{A} \leftrightarrow A$ in T. Therefore it suffices to prove the assertion for $\neg, \wedge$-formulae $\widetilde{A}$ by induction on the F-rank.

I. $\widetilde{A} \equiv (r \overset{o}{=} s)$. By (T22),(S): $r \overset{o}{=} s \leftrightarrow \varphi_{\widetilde{A}} \underline{\underline{u}} \; \equiv: \; +(-rs)(-sr) = 0$

II. $\widetilde{A} \equiv \neg \widetilde{B}$ or $\widetilde{A} \equiv \widetilde{B} \wedge \widetilde{C}$

By induction hypothesis: $\varphi_{\widetilde{B}} \underline{\underline{u}} = 0 \leftrightarrow \widetilde{B}$, $\varphi_{\widetilde{C}} \underline{\underline{u}} = 0 \leftrightarrow \widetilde{C}$

$\widetilde{A} \equiv \neg \widetilde{B} \leftrightarrow \varphi_{\widetilde{B}} \underline{\underline{u}} \neq 0 \leftrightarrow \varphi_{\widetilde{A}} \underline{\underline{u}} \; \equiv: \; \overline{\overline{sg}}(\varphi_{\widetilde{B}} \underline{\underline{u}}) = 0$ (T23)

$\widetilde{A} \equiv (\widetilde{B} \wedge \widetilde{C}) \leftrightarrow \varphi_{\widetilde{B}} \underline{\underline{u}} = 0 \wedge \varphi_{\widetilde{C}} \underline{\underline{u}} = 0 \leftrightarrow \varphi_{\widetilde{A}} \underline{\underline{u}} \; \equiv: \; +(\varphi_{\widetilde{B}} \underline{\underline{u}})(\varphi_{\widetilde{C}} \underline{\underline{u}}) = 0$ (T12)

(T25) T is identical with the calculus of the primitive recursive functionals. - (1.1), (4.1) are valid for T and all extensions.

Proof: (T21),(Mp); (S); (T5),(T0);(T6),(T7),(CI); (5.2),(T4); (E).

(T26) For preceding functionals φ_1, φ_2, ψ there exists a functional φ in T with: $\psi \underline{\underline{u}} \overset{o}{=} 0 \rightarrow \varphi \underline{\underline{u}} \overset{a}{=} \varphi_1 \underline{\underline{u}}$, $\psi \underline{\underline{u}} \neq 0 \rightarrow \varphi \underline{\underline{u}} \overset{a}{=} \varphi_2 \underline{\underline{u}}$.

Proof:

$\chi 0 = \varphi_1$, $\chi u' = \varphi_2$ (T4)

$\varphi \underline{\underline{u}} \overset{a}{=} \chi(\psi \underline{\underline{u}}) \underline{\underline{u}}$ (5.2)

Now: $\psi \underline{\underline{u}} \overset{o}{=} 0 \rightarrow \varphi \underline{\underline{u}} \overset{a}{=} \chi 0 \underline{\underline{u}} = \varphi_1 \underline{\underline{u}}$

$\quad \psi \underline{\underline{u}} \neq 0 \rightarrow \psi \underline{\underline{u}} \overset{o}{=} (\delta(\psi \underline{\underline{u}}))'$ (T8)

$\quad \quad \rightarrow \varphi \underline{\underline{u}} \overset{a}{=} \chi((\delta(\psi \underline{\underline{u}}))') \underline{\underline{u}} = \varphi_2 \underline{\underline{u}}$

(T27) For preceding functionals φ_1, φ_2 and a quantifierfree formula $A(\underline{\underline{u}})$, containing only preceding functionals and at most the variables $\underline{\underline{u}}$, there is a functional φ in T with: $A(\underline{\underline{u}}) \rightarrow \varphi \underline{\underline{u}} \overset{a}{=} \varphi_1 \underline{\underline{u}}$, $\neg A(\underline{\underline{u}}) \rightarrow \varphi \underline{\underline{u}} \overset{a}{=} \varphi_2 \underline{\underline{u}}$.

Proof: (T24),(T26)

(T28) $-uu = 0$, i.e. $u \leq u$, $\neg(u<u)$.

Proof: Induction on u

I. $-00 = 0$

II. $-uu=0 \longrightarrow -u'u'=-uu=0$ (T14)

(T29) $u \leq v \longleftrightarrow u \overset{o}{=} v \vee u<v$

Proof:

1.$\rightarrow$: $\underbrace{-vu=0}_{u\leq v} \wedge -uv=0 \longrightarrow +(-uv)(-vu)=0$

$\longrightarrow u \overset{o}{=} v$ (T22)

With (T21): $u \leq v \longrightarrow u=v \vee \underbrace{-uv \neq 0}_{u<v}$

2.$\leftarrow$: $u \overset{o}{=} v \longrightarrow -vu=-uu=0 \longleftrightarrow u \leq v$ (T28)

$\underbrace{-uv \neq 0}_{u<v} \longrightarrow v=+u(-uv)$ (T15)

(*) $\longrightarrow -vu=-(+u(-uv))u$

By induction on u one proves $-(+uv)u = 0$:

I. $-(+0v)0 = -v0 = 0$ follows by induction on v, because

1. $-00=0$ 2. $-v0=0 \longrightarrow -v'0=\delta(-v0)=\delta 0=0$

II. $-(+uv)u=0 \longrightarrow -(+u'v)u'=-(+uv)'u'=-(+uv)u=0$ (T10),(T14)

Together with (*) this yields $u<v \longrightarrow -vu=0 \longleftrightarrow u \leq v$.

(T30) $+(+uv)w = +u(+vw)$

Proof: Induction on u

I. $+(+0v)w = +vw = +0(+vw)$

II. $+(+uv)w=+u(+vw) \longrightarrow +(+u'v)w=(+(+uv)w)'=(+u(+vw))'=+u'(+vw)$ (T10)

(T31) $u<v \wedge v<w \longrightarrow u<w$

Proof:

$\underbrace{u<v}_{-uv\neq 0} \wedge v<w \wedge \underbrace{-uw=0}_{u \geq w} \longrightarrow v=+u(-uv) \wedge w=+v(-vw) \wedge u=+w(-wu) \wedge -uv=(\delta(-uv))'$ (T15),(T17),(T8)

$\longrightarrow u=+(+(+u(-uv))(-vw))(-wu) \wedge -uv=(\delta(-uv))'$

$\longrightarrow u=+u'(+(+\delta(-uv)(-vw))(-wu))$ (T30),(T10)

$\longrightarrow \curlywedge$

because induction on u gives $+u'v \neq u$:

I. $+0'v = (+0v)' \neq 0$ (T6)

II. $+u'v \neq u \longrightarrow +u''v=(+u'v)' \neq u'$ (T7)

Now (T21),(T13).

(T32) $v<u \longleftrightarrow -vu=(-v'u)'$

<u>Proof:</u>

$-vu=(-v'u)' \longrightarrow -vu\neq 0 \longleftrightarrow v<u$ (T6)

$v<u \longleftrightarrow -vu\neq 0 \longrightarrow -vu=(\delta(-vu))'=(-v'u)'$ (T8)

(T33) $u\neq 0 \longleftrightarrow -(\delta u)u=1$

<u>Proof:</u>

1.$\leftarrow$: $-(\delta u)u=1 \wedge u=0 \longrightarrow 1=-(\delta u)u=-00=0$

2.$\rightarrow$: Induction on u

I. $0\neq 0 \rightarrow A$

II. $u\neq 0 \longrightarrow (u\neq 0 \rightarrow -(\delta u)u=1) \wedge u'\neq 0 \longrightarrow -(\delta u')u'=-uu'=-(\delta u)'u'=-(\delta u)u=1$ (T8),(T14)

$u=0 \longrightarrow -(\delta u')u'=-uu'=-01=1$

The induction step results from the last two rows by (T9).

(T34) $w'<u \longrightarrow -(-w'u)u=w'$

<u>Proof:</u> Induction on w

I. $0'<u \rightarrow u\neq 0$ (T16),(T31),(T28)

$\longrightarrow -(-0'u)u=-(\delta u)u=0'$ (T33)

II. $\{w'<u \rightarrow -(-w'u)u=w'\} \wedge w''<u \longrightarrow w'<u$ (T16),(T31)

$\longrightarrow 0\neq w'=-(-w'u)u=-(-w''u)'u=\delta(-(-w''u)u)$ (T6),(T32)

$\longrightarrow 0\neq -(-w''u)u=(\delta(-(-w''u)u))'=w''$ (T8)

(T35) $v\leqslant u \longrightarrow v=-(-vu)u$

<u>Proof:</u>

$v=0 \longrightarrow -(-vu)u=-uu=0=v$ (T28)

$v=u \longrightarrow -(-vu)u=-0u=u=v$ (T28)

$v\neq 0 \wedge v<u \longrightarrow v=(\delta v)' \wedge (\delta v)'<u$ (T8)

$\longrightarrow -(-vu)u=-(-(\delta v)'u)u=(\delta v)'=v$ (T34)

Now (T35) by (T21) and (T29).

(T36) The rule $\dfrac{A(0,w^{\alpha}),\ A(v^{0},tvw^{\alpha}) \rightarrow A(v',w)}{A(v,w)}$ $(v \notin A(0,w),t)$ is explicit derivable in T.

<u>Proof:</u>

$\underline{\underline{u}}_1$ be the stock of variables in t; $v^{0}\notin\underline{\underline{u}}_1$.

$\psi\underline{\underline{u}}_1 0 \overset{\alpha o}{=} \lambda x^o, y^\alpha \cdot y$

$\psi\underline{\underline{u}}_1 v' = \lambda x^o, y^\alpha \cdot t(-v'x)(\psi\underline{\underline{u}}_1 vxy)$ (T4),(5.2)

$(*)\ v<u \longrightarrow \psi\underline{\underline{u}}_1(-vu)uw \overset{\alpha}{=} \psi\underline{\underline{u}}_1(-v'u)'uw = t(-(-v'u)'u)(\psi\underline{\underline{u}}_1(-v'u)uw)$

$= t(-(-vu)u)(\psi\underline{\underline{u}}_1(-v'u)uw) = tv(\psi\underline{\underline{u}}_1(-v'u)uw)$ (T32),(T35),(T29)

Take $B(v^o) \equiv: (v \leqslant u \rightarrow A(v, \psi\underline{\underline{u}}_1(-vu)uw))$. By induction on v B(v) is provable:

I. $B(0) \equiv: (0 \leqslant u \rightarrow A(0, \psi\underline{\underline{u}}_1(-0u)uw))$ follows from the premise with (S).

II. $(v \leqslant u \rightarrow A(v, \psi\underline{\underline{u}}_1(-vu)uw)) \wedge v' \leqslant u \longrightarrow v<u$ (T16),(T29),(T31)

$\longrightarrow A(v, tv(\psi\underline{\underline{u}}_1(-v'u)uw))$ (T29),(*),(E)$^\alpha$

$\longrightarrow A(v', \psi\underline{\underline{u}}_1(-v'u)uw)$ premise,(S)

Especially $v \leqslant v \rightarrow A(v, \psi\underline{\underline{u}}_1(-vv)vw)$. With (T28),(E) now A(v,w).

(T37) $u \geqslant 0,\ u' > 0,\ \neg(u < 0)$

Proof: Induction on u with (T28),(T16),(T31),(T29),(T6).

(T38) $u<v \leftrightarrow u' \leqslant v,\ u \leqslant v \leftrightarrow u' \leqslant v' \leftrightarrow u<v'$ (T14)

Proof:

1.$\leftarrow$: (T29),(T16),(T31)

2.$\rightarrow$: Induction on u gives: (a) $\delta u = 0 \longrightarrow u=0 \vee u=1$ (induction step: $\delta u' = 0 \longrightarrow u = \delta u' = 0 \longrightarrow u'=0 \vee u'=1$)

$\underset{\smile}{u<v} \wedge -u'v=0 \longrightarrow 0 = -u'v = \delta(-uv) \wedge -uv \neq 0$

$-uv \neq 0 \quad \longrightarrow -uv = 0' \neq 0 \wedge u<v$ (a),(T6)

$\longrightarrow v = +u(-uv) = +u0' = u'$ (T15),(T10),(T11)

With (T21),(T29): $u<v \longrightarrow u'=v \vee -u'v \neq 0 \leftrightarrow u' \leqslant v$

(T39) There is a functional <> in T with:

$w_1 > w_2 \longrightarrow \langle\rangle u^{\alpha o} v^{\alpha o} w_1^o w_2^o \overset{\alpha}{=} uw_2,\quad w_1 \leqslant w_2 \longrightarrow \langle\rangle uvw_1w_2 \overset{\alpha}{=} vw_2$

$\langle\rangle uvw_1$ corresponds to the informal $\langle u0, \ldots, u(w_1-1), vw_1, v(w_1'), \ldots\rangle$.

Proof: $\neg(w_1 > w_2) \leftrightarrow w_1 \leqslant w_2$ by (T13),(T29),(T28),(T31). Now (T26),(5.2).

$(\sigma)\ \sigma^o \overset{o}{=} 0,\quad \sigma^{\alpha\beta}u^\beta \overset{\alpha}{=} \sigma^\alpha$ (zero-functionals)

(T40) $\overline{r^{\alpha o}, u^o} \overset{\alpha o}{=} \langle\rangle r\sigma u,\quad v^o < u \longrightarrow \overline{r,u}\, v \overset{\alpha}{=} rv,\quad v \geqslant u \longrightarrow \overline{r,u}\, v \overset{\alpha}{=} \sigma,\quad \overline{r,0}\, v \overset{\alpha}{=} \sigma$

$\overline{r,u}$ corresponds to the informal <r0,...,r(u-1)>.

Proof: (T39),(T37)

(T41) $\overline{r^{\alpha o},u^{o}}*t^{\alpha} \overset{\alpha o}{=} <>(<>r(\lambda x^{o}t)u)\sigma u'$, $v<u \longrightarrow (\overline{r,u}*t)v \overset{\alpha}{=} rv$, $(\overline{r,u}*t)u \overset{\alpha}{=} t$, $v \geqslant u' \longrightarrow (\overline{r,u}*t)v \overset{\alpha}{=} \sigma$

$\overline{r,u}*t$ corresponds to the informal <r0,...,r(u-1),t>.

Proof: (T39),(T16),(T31),(T29)

(T42) (Simultaneous recursion)

For preceding functionals $\psi_1,\chi_1,\ldots,\psi_n,\chi_n$ there are functionals $\varphi_1,\ldots,\varphi_n$ in T such that intensionally (i.e. without (E), but with the equality axioms for $\overset{\alpha}{=}$):

$$\left.\begin{array}{l} \varphi_i^{\alpha_i o}0 \overset{\alpha_i}{=} \psi_i \\ \varphi_i u' = \chi_i(\varphi_1 u)\ldots(\varphi_n u)u^{o} \end{array}\right\} \quad i=1,\ldots,n$$

Proof: Metainduction on n

I. n=1: (T4)

II. By induction hypothesis there are $\tilde{\varphi}_1,\ldots,\tilde{\varphi}_n$ in T so that for i=1,...,n:

(1) $\tilde{\varphi}_i 0 = \lambda x^{\alpha_{n+1}o}\psi_i$, $\tilde{\varphi}_i u' = \lambda x^{\alpha_{n+1}o}\chi_i(\tilde{\varphi}_1 ux)\ldots(\tilde{\varphi}_n ux)(xu)u^{o}$

By (T4),(5.2):

(2) $$\begin{array}{l} \hat{\varphi}_{n+1}^{\alpha_{n+1}o}0 = \overline{\lambda y^{o}\psi_{n+1},1} \\ \hat{\varphi}_{n+1}u' \overset{\alpha_{n+1}o}{=} \overline{\hat{\varphi}_{n+1}u,u'}*(\chi_{n+1}(\tilde{\varphi}_1 u(\hat{\varphi}_{n+1}u))\ldots(\tilde{\varphi}_n u(\hat{\varphi}_{n+1}u))(\hat{\varphi}_{n+1}uu)u) \end{array}$$

(3) $\varphi_{n+1}^{\alpha_{n+1}o}u^{o} =: \hat{\varphi}_{n+1}uu$

(4) $\varphi_i^{\alpha_i o}u =: \tilde{\varphi}_i u\varphi_{n+1}$ i=1,...,n

For these φ_j (j=1,...,n+1) the simultaneous recursion equations above will be proved in T intensionally. - For i=1,...,n:

$\varphi_i 0 = \tilde{\varphi}_i 0\varphi_{n+1} = \psi_i$ (4),(1)

$\varphi_i u' = \tilde{\varphi}_i u'\varphi_{n+1} = \chi_i(\tilde{\varphi}_1 u\varphi_{n+1})\ldots(\tilde{\varphi}_n u\varphi_{n+1})(\varphi_{n+1}u)u$ (4),(1)

$= \chi_i(\varphi_1 u)\ldots(\varphi_n u)(\varphi_{n+1}u)u$ (4)

$\varphi_{n+1}0 = \hat{\varphi}_{n+1}00 = \overline{\lambda y^{o}\psi_{n+1},1}0 = \psi_{n+1}$ (3),(2),(T40),(T16)

$\varphi_{n+1}u' = \hat{\varphi}_{n+1}u'u' = \chi_{n+1}(\tilde{\varphi}_1 u(\hat{\varphi}_{n+1}u))\ldots(\tilde{\varphi}_n u(\hat{\varphi}_{n+1}u))(\varphi_{n+1}u)u$ (3),(2),(T41)

It remains to show: $\bigwedge_{i=1}^{n} \tilde{\varphi}_i u(\hat{\varphi}_{n+1}u)=\varphi_i u=\tilde{\varphi}_i u\varphi_{n+1}$ (4). Induction on u:

I. $\tilde{\varphi}_i 0(\hat{\varphi}_{n+1}0)=\psi_i=\tilde{\varphi}_i 0\varphi_{n+1}$ (1)

II. $\bigwedge_{i=1}^{n} \tilde{\varphi}_i u(\hat{\varphi}_{n+1}u)=\tilde{\varphi}_i u\varphi_{n+1} \longrightarrow \tilde{\varphi}_i u'\varphi_{n+1}=\chi_i(\tilde{\varphi}_1 u\varphi_{n+1})\dots(\tilde{\varphi}_n u\varphi_{n+1})(\varphi_{n+1}u)u$ (1)

$=\chi_i(\tilde{\varphi}_1 u(\hat{\varphi}_{n+1}u))\dots(\tilde{\varphi}_n u(\hat{\varphi}_{n+1}u))(\varphi_{n+1}u)u$

$\wedge\ \tilde{\varphi}_i u'(\hat{\varphi}_{n+1}u')=\chi_i(\tilde{\varphi}_1 u(\hat{\varphi}_{n+1}u'))\dots(\tilde{\varphi}_n u(\hat{\varphi}_{n+1}u'))(\hat{\varphi}_{n+1}u'u)u$ (1)

As $\hat{\varphi}_{n+1}u'u=\hat{\varphi}_{n+1}uu=\varphi_{n+1}u$ (2),(T41),(T16),(3) , so it suffices for the induction step to prove $\bigwedge_{i=1}^{n} \tilde{\varphi}_i u(\hat{\varphi}_{n+1}u')=\tilde{\varphi}_i u(\hat{\varphi}_{n+1}u)$. More generally (T28) we show $\bigwedge_{i=1}^{n} \{w\leq u \longrightarrow \tilde{\varphi}_i w(\hat{\varphi}_{n+1}u)=\tilde{\varphi}_i w(\hat{\varphi}_{n+1}u')\}$ by induction on w:

1. $\tilde{\varphi}_i 0(\hat{\varphi}_{n+1}u)=\psi_i=\tilde{\varphi}_i 0(\hat{\varphi}_{n+1}u')$ (1)

2. (5) $w\leq u \longrightarrow w<u' \wedge \hat{\varphi}_{n+1}u'w=\hat{\varphi}_{n+1}uw$ (T38),(2),(T41)

$V :\equiv \bigwedge_{i=1}^{n} \{w\leq u \longrightarrow \tilde{\varphi}_i w(\hat{\varphi}_{n+1}u)=\tilde{\varphi}_i w(\hat{\varphi}_{n+1}u')\} \wedge w'\leq u$

$V \longrightarrow w\leq u \wedge \tilde{\varphi}_i w'(\hat{\varphi}_{n+1}u)=\chi_i(\tilde{\varphi}_1 w(\hat{\varphi}_{n+1}u))\dots(\tilde{\varphi}_n w(\hat{\varphi}_{n+1}u))(\hat{\varphi}_{n+1}uw)w$

$=\chi_i(\tilde{\varphi}_1 w(\hat{\varphi}_{n+1}u'))\dots(\tilde{\varphi}_n w(\hat{\varphi}_{n+1}u'))(\hat{\varphi}_{n+1}u'w)w=\tilde{\varphi}_i w'(\hat{\varphi}_{n+1}u')$

(T38),(T29),(1), premise V,(5)

VI. Functional interpretation of classical arithmetic plus (ER)-qf, (AC)-qf and functional interpretation in the narrower sense of Heyting-analysis plus (ER)-qf, (MP), ($\underset{\rightarrow}{V}$) in T.

To get a functional interpretation of classical arithmetic in T we have to complete (3.8) by the solution in T of the functional interpretation in the narrower sense for Heyting-arithmetic. We show that this holds for the axioms and is hereditary for the rules. Using (4.1), for the formulae-signs A, B, C in the instructions for the axioms and rules we put in $A' \equiv \bigvee x_1 \bigwedge y_1 A_o(x_1, y_1, \underline{u}_1)$, $B' \equiv \bigvee x_2 \bigwedge y_2 B_o(x_2, y_2, \underline{u}_2)$, $C' \equiv \bigvee x_3 \bigwedge y_3 C_o(x_3, y_3, \underline{u}_3)$ with quantifierfree A_o, B_o, C_o out of T. Here $\underline{u}_i$ are the single stocks of free variables; $\underline{u} \equiv: \underline{u}_1 \cup \underline{u}_2 \cup \underline{u}_3$. If quantifiers are missing all magnitudes referring to these arguments have to be crossed out. - For simplicity we write A' instead of var(A'). Formally we also abbreviate by relating the $\underline{\varphi}$-solution of $\bigvee x_1, .., x_m \bigwedge y_1, .., y_n A_o(x_1, .., x_m, y_1, .., y_n, \underline{u})$ in a direct way to the indices used, i.e. in taking $A(\varphi_1 \underline{u}, .., \varphi_m \underline{u}, v_1, .., v_n, \underline{u})$.

Finally we include (ER)-qf, (AC), (MP) and ($\underset{\rightarrow}{V}$). - With more expense (compare chapter I) all this - (ER)-qf of course excepted - can be done without qct in a similar manner entirely intensional.

(6.1) If A (out of the functional language) is a law of intuitionistic predicate logic with equality (I1,3) then there is a functional φ in T with: ${}_T\vdash \varphi\text{-sol}(A')$.

Proof: Induction on the length of proof

I. Intuitionistic predicate logic

A. Axioms

1. (Taut)

a) $(A \vee A \rightarrow A)' \equiv \bigvee x_1, x_2, x_3 \bigwedge y_1^o, .., y_4 \{(y_1 = 0 \wedge A_o(y_2, x_2\underline{y}, \underline{u})) \vee$

$$(y_1 \neq 0 \wedge A_o(y_3, x_3\underline{y}, \underline{u})) \longrightarrow A_o(x_1 y_1 y_2 y_3, y_4, \underline{u})\}$$

$\underline{\varphi}$-solution in T: $\varphi_1 \underline{u} 0 v_2 v_3 = v_2$, $\varphi_1 \underline{u} v_1' v_2 v_3 = v_3$ (T4),(5.2)

$\varphi_2 \underline{u}\underline{v} = v_4$, $\varphi_3 \underline{u}\underline{v} = v_4$ (5.2)

b) $(A \to A \wedge A)' \equiv \bigvee x_1,x_2,x_3 \bigwedge y_1,y_2,y_3\{A_o(y_1,x_3\underline{y},\underline{u}) \to A_o(x_1y_1,y_2,\underline{u}) \wedge$

$A_o(x_2y_1,y_3,\underline{u})\}$

φ-solution in T: $\varphi_1\underline{u}v_1 = \varphi_2\underline{u}v_1 = v_1$ (5.2)

$$\varphi_3\underline{u}\underline{v} = \begin{cases} v_3 & \text{if } A_o(v_1,v_2,\underline{u}) \\ v_2 & \text{if } \neg A_o(v_1,v_2,\underline{u}) \end{cases} \qquad \text{(T27),(5.2)}$$

2. (Add)

a) $(A \to A \vee B)' \equiv \bigvee x_1,\ldots,x_4 \bigwedge y_1,y_2,y_3\{A_o(y_1,x_4\underline{y},\underline{u}_1) \to$

$(x_1y_1=0 \wedge A_o(x_2y_1,y_2,\underline{u}_1)) \vee (x_1y_1 \neq 0 \wedge B_o(x_3y_1,y_3,\underline{u}_2))\}$

φ-solution in T: $\varphi_1\underline{u}v_1 \overset{\circ}{=} 0,\ \ \varphi_2\underline{u}v_1 = v_1$

$\varphi_3\underline{u}v_1 = \sigma,\ \ \varphi_4\underline{u}\underline{v} = v_2$ (5.2)

b) $(A \wedge B \to A)' \equiv \bigvee x_1,x_2,x_3 \bigwedge y_1,y_2,y_3\{A_o(y_1,x_2\underline{y},\underline{u}_1) \wedge B_o(y_2,x_3\underline{y},\underline{u}_2) \to$

$A_o(x_1y_1y_2,y_3,\underline{u}_1)\}$

φ-solution in T: $\varphi_1\underline{u}v_1v_2 = v_1,\ \ \varphi_2\underline{u}\underline{v} = v_3,\ \ \varphi_3\underline{u}\underline{v} = \sigma$ (5.2)

3. (Perm)

a) $(A \vee B \to B \vee A)' \equiv \bigvee x_1,\ldots,x_5 \bigwedge y_1^{\circ},\ldots,y_5\{(y_1=0 \wedge A_o(y_2,x_4\underline{y},\underline{u}_1)) \vee (y_1 \neq 0 \wedge$
$B_o(y_3,x_5\underline{y},\underline{u}_2)) \to (x_1y_1y_2y_3=0 \wedge B_o(x_2y_1y_2y_3,y_4,\underline{u}_2)) \vee (x_1y_1y_2y_3 \neq 0 \wedge$

$A_o(x_3y_1y_2y_3,y_5,\underline{u}_1))\}$

φ-solution in T: $\varphi_1\underline{u}v_1^{\circ}v_2v_3 = \overline{\overline{sg}}(v_1),\ \ \varphi_2\underline{u}v_1v_2v_3 = v_3$

$\varphi_3\underline{u}v_1v_2v_3 = v_2,\ \ \varphi_4\underline{u}\underline{v} = v_5,\ \ \varphi_5\underline{u}\underline{v} = v_4$ (5.2)

b) $(A \wedge B \to B \wedge A)' \equiv \bigvee x_1,\ldots,x_4 \bigwedge y_1,\ldots,y_4\{A_o(y_1,x_3\underline{y},\underline{u}_1) \wedge B_o(y_2,x_4\underline{y},\underline{u}_2) \to$
$B_o(x_1y_1y_2,y_3,\underline{u}_2) \wedge A_o(x_2y_1y_2,y_4,\underline{u}_1)\}$

φ-solution in T: $\varphi_1\underline{u}v_1v_2 = v_2,\ \ \varphi_2\underline{u}v_1v_2 = v_1$

$\varphi_3\underline{u}\underline{v} = v_4,\ \ \varphi_4\underline{u}\underline{v} = v_3$ (5.2)

4. $(\perp)$ $(\perp \to A)' \equiv \bigvee x_1 \bigwedge y_1(\perp \to A(x_1,y_1))$

φ-solution in T: $\varphi = \sigma$

5. (Q)

a) $(\bigwedge xA(x) \rightarrow A(r))' \equiv \bigvee x_1,x_2,x_3 \bigwedge y_1,y_2\{A_o(y_1(x_2\underline{y}),x_3\underline{y},x_2\underline{y},\underline{\underline{u}}) \longrightarrow$

$A_o(x_1y_1,y_2,r(\underline{\underline{u}}),\underline{\underline{u}})\}$

$\underline{\varphi}$-solution in T: $\varphi_1\underline{\underline{u}}v_1 = v_1(r(\underline{\underline{u}})),\quad \varphi_2\underline{\underline{u}}\underline{v} = r(\underline{\underline{u}}),\quad \varphi_3\underline{\underline{u}}\underline{v} = v_2$ (5.2)

b) $(A(r) \rightarrow \bigvee xA(x))' \equiv \bigvee x_1,x_2,x_3 \bigwedge y_1,y_2\{A_o(y_1,x_3\underline{y},r(\underline{\underline{u}}),\underline{\underline{u}}) \longrightarrow$

$A_o(x_2y_1,y_2,x_1y_1,\underline{\underline{u}})\}$

$\underline{\varphi}$-solution in T: $\varphi_1\underline{\underline{u}}v_1 = r(\underline{\underline{u}}),\quad \varphi_2\underline{\underline{u}}v_1 = v_1,\quad \varphi_3\underline{\underline{u}}\underline{v} = v_2$ (5.2)

B. Rules

1. (Mp)

Given are the φ-solution of $A' \equiv \bigvee x \bigwedge yA_o(x,y,\underline{\underline{u}}_1)$ and the $\underline{\psi}$-solution of $(A \rightarrow B)' \equiv \bigvee x_1,x_2 \bigwedge y_1,y_2\{A_o(y_1,x_2\underline{y},\underline{\underline{u}}_1) \longrightarrow B_o(x_1y_1,y_2,\underline{\underline{u}}_2)\}$ in T:

$A_o(\varphi\underline{\underline{u}}_1,v,\underline{\underline{u}}_1),\quad A_o(w_1,\psi_2\underline{\underline{u}}\underline{w},\underline{\underline{u}}_1) \longrightarrow B_o(\psi_1\underline{\underline{u}}w_1,w_2,\underline{\underline{u}}_2),\quad \underline{w} \equiv w_1w_2$

From this by the substitution $w_1\colon \varphi\underline{\underline{u}}_1,\ v\colon \psi_2\underline{\underline{u}}(\varphi\underline{\underline{u}}_1)w_2$ one gets $B_o(\psi_1\underline{\underline{u}}(\varphi\underline{\underline{u}}_1),w_2,\underline{\underline{u}}_2)$. Under $\underline{\underline{u}}_1 \equiv \underline{\underline{u}}_o \cup \hat{\underline{\underline{u}}}_2$ ($\underline{\underline{u}}_o$ disjoint to $\underline{\underline{u}}_2$, $\hat{\underline{\underline{u}}}_2$ part of $\underline{\underline{u}}_2$), $\underline{\underline{u}} \equiv \underline{\underline{u}}_o \cup \underline{\underline{u}}_2$ therefore $\chi\underline{\underline{u}}_2 = \psi_1\underline{\underline{\sigma}}\underline{\underline{u}}_2(\varphi\underline{\underline{\sigma}}\cup\hat{\underline{\underline{u}}}_2)$ according to (5.2) is a solution of $B' \equiv \bigvee x_2 \bigwedge y_2B_o(x_2,y_2,\underline{\underline{u}}_2)$ in T.

2. (Syll)

Given are the $\underline{\varphi}$-solution of $(A \rightarrow B)' \equiv \bigvee x_1,x_2 \bigwedge y_1,y_2\{A_o(y_1,x_2\underline{y},\underline{\underline{u}}_1) \longrightarrow B_o(x_1y_1,y_2,\underline{\underline{u}}_2)\}$ and the $\underline{\psi}$-solution of $(B \rightarrow C)' \equiv \bigvee x_1,x_2 \bigwedge z_1,z_2$ $\{B_o(z_1,x_2\underline{z},\underline{\underline{u}}_2) \longrightarrow C_o(x_1z_1,z_2,\underline{\underline{u}}_3)\}$ in T:

$A_o(v_1,\varphi_2\underline{\underline{u}}_1\underline{\underline{u}}_2\underline{v},\underline{\underline{u}}_1) \longrightarrow B_o(\varphi_1\underline{\underline{u}}_1\underline{\underline{u}}_2v_1,v_2,\underline{\underline{u}}_2),\quad B_o(w_1,\psi_2\underline{\underline{u}}_2\underline{\underline{u}}_3\underline{w},\underline{\underline{u}}_2) \longrightarrow$

$C_o(\psi_1\underline{\underline{u}}_2\underline{\underline{u}}_3w_1,w_2,\underline{\underline{u}}_3)$

The substitution $w_1\colon \varphi_1\underline{\underline{u}}_1\underline{\underline{u}}_2v_1,\ v_2\colon \psi_2\underline{\underline{u}}_2\underline{\underline{u}}_3(\varphi_1\underline{\underline{u}}_1\underline{\underline{u}}_2v_1)w_2$ gives

$A_o(v_1,\varphi_2\underline{\underline{u}}_1\underline{\underline{u}}_2v_1(\psi_2\underline{\underline{u}}_2\underline{\underline{u}}_3(\varphi_1\underline{\underline{u}}_1\underline{\underline{u}}_2v_1)w_2),\underline{\underline{u}}_1) \longrightarrow C_o(\psi_1\underline{\underline{u}}_2\underline{\underline{u}}_3(\varphi_1\underline{\underline{u}}_1\underline{\underline{u}}_2v_1),w_2,\underline{\underline{u}}_3)$

Under $\underline{\underline{u}}_2 \equiv \underline{\underline{u}}_o \cup \hat{\underline{\underline{u}}}_1 \cup \hat{\underline{\underline{u}}}_3$ ($\underline{\underline{u}}_o$ disjoint to $\underline{\underline{u}}_1,\underline{\underline{u}}_3$; $\hat{\underline{\underline{u}}}_1,\hat{\underline{\underline{u}}}_3$ parts of $\underline{\underline{u}}_1$ resp. $\underline{\underline{u}}_3$)

therefore $\chi_1\underline{\underline{u}}_1\underline{\underline{u}}_3v_1 = \psi_1\underline{\underline{\sigma}}\cup\hat{\underline{\underline{u}}}_1\cup\hat{\underline{\underline{u}}}_3\ \underline{\underline{u}}_3(\varphi_1\underline{\underline{u}}_1\ \underline{\underline{\sigma}}\cup\hat{\underline{\underline{u}}}_1\cup\hat{\underline{\underline{u}}}_3\ v_1)$,

$\chi_2\underline{u}_1\underline{u}_3 v_1 w_2 = \varphi_2\underline{u}_1\ \underline{\sigma}\underline{v}\hat{\underline{u}}_1\underline{v}\hat{\underline{u}}_3\ v_1(\psi_2\underline{\sigma}\underline{v}\hat{\underline{u}}_1\underline{v}\hat{\underline{u}}_3\ \underline{u}_3(\varphi_1\underline{u}_1\ \underline{\sigma}\underline{v}\hat{\underline{u}}_1\underline{v}\hat{\underline{u}}_3\ v_1)w_2)$

according to (5.2) are a χ-solution of $(A\rightarrow C)' \equiv \bigvee x_1,x_2 \bigwedge y_1,z_2 \{A_o(y_1,x_2y_1z_2,\underline{u}_1) \longrightarrow B_o(x_1y_1,z_2,\underline{u}_3)\}$ in T.

3. (Sum)

Given is the φ-solution of $(A\rightarrow B)' \equiv \bigvee x_1,x_2 \bigwedge y_1,y_2\{A_o(y_1,x_2\underline{y},\underline{u}_1) \longrightarrow B_o(x_1y_1,y_2,\underline{u}_2)\}$ in T: $A_o(v_1,\varphi_2\underline{u}_1\underline{u}_2\underline{v},\underline{u}_1) \longrightarrow B_o(\varphi_1\underline{u}_1\underline{u}_2v_1,v_2,\underline{u}_2)$

We have to produce a ψ-solution of $(C \vee A \rightarrow C \vee B)' \equiv \bigvee x_1,\ldots,x_5 \bigwedge y_1^o,\ldots,y_5$ $\{(y_1=0 \wedge C_o(y_2,x_4\underline{y},\underline{u}_3)) \vee (y_1 \neq 0 \wedge A_o(y_3,x_5\underline{y},\underline{u}_1)) \longrightarrow$ $(x_1y_1y_2y_3=0 \wedge C_o(x_2y_1y_2y_3,y_4,\underline{u}_3)) \vee (x_1y_1y_2y_3 \neq 0 \wedge B_o(x_3y_1y_2y_3,y_5,\underline{u}_2))\}$

in T. According to (5.2) this is performed by the following functionals:

$\psi_1\underline{u}v_1v_2v_3 \overset{o}{=} v_1,\quad \psi_2\underline{u}v_1v_2v_3 = v_2$

$\psi_3\underline{u}v_1v_2v_3 = \varphi_1\underline{u}_1\underline{u}_2v_3,\quad \psi_4\underline{u}\underline{v} = v_4,\quad \psi_5\underline{u}\underline{v} = \varphi_2\underline{u}_1\underline{u}_2v_3v_5$

4. (Exp), (Imp)

As $(A \wedge B \rightarrow C)' \equiv \bigvee x_1,x_2,x_3 \bigwedge y_1,y_2,y_3\{A_o(y_1,x_2\underline{y},\underline{u}_1) \wedge B_o(y_2,x_3\underline{y},\underline{u}_2) \longrightarrow C_o(x_1y_1y_2,y_3,\underline{u}_3)\}$ and $(A\rightarrow B \rightarrow C)' \equiv \bigvee x_1,x_2,x_3 \bigwedge y_1,y_2,y_3\{A_o(y_1,x_2\underline{y},\underline{u}_1) \longrightarrow B_o(y_2,x_3\underline{y},\underline{u}_2) \rightarrow C_o(x_1y_1y_2,y_3,\underline{u}_3)\}$, so every solution of $(A \wedge B \rightarrow C)'$ in T is also a T-solution of $(A \rightarrow B \rightarrow C)'$ and vice versa.

5. ($\bigwedge$)

As $(A \rightarrow B(u))' \equiv \bigvee x_1,x_2 \bigwedge y_1,y_2\{A_o(y_1,x_2\underline{y},\underline{u}_1) \longrightarrow B_o(x_1y_1,y_2,u,\underline{u}_2)\}$, $u \notin \underline{u}_1,\underline{u}_2$ and $(A\rightarrow \bigwedge x B(x))' \equiv \bigvee x_1,x_2 \bigwedge y_o,y_1,y_2\{A_o(y_1,x_2y_o\underline{y},\underline{u}_1) \longrightarrow B_o(x_1y_oy_1,y_2,y_o,\underline{u}_2)\}$, so arises in T from each solution of $(A\rightarrow B(u))'$ by identification of u with v_o a solution of $(A\rightarrow \bigwedge x B(x))'$ and vice versa.

6. ($\bigvee$)

Since $(A(u)\rightarrow B)' \equiv \bigvee x_1,x_2 \bigwedge y_1,y_2\{A_o(y_1,x_2\underline{y},u,\underline{u}_1) \rightarrow B_o(x_1y_1,y_2,\underline{u}_2)\}$, $u \notin \underline{u}_1,\underline{u}_2$ and $(\bigvee x A(x) \rightarrow B)' \equiv \bigvee x_1,x_2 \bigwedge y_o,y_1,y_2\{A_o(y_1,x_2y_o\underline{y},y_o,\underline{u}_1) \longrightarrow B_o(x_1y_oy_1,y_2,\underline{u}_2)\}$ so in T also here from each solution of $(A(u)\rightarrow B)'$ by identification of u with v_o a solution of $(\bigvee x A(x) \rightarrow B)'$ originates and vice versa.

II. Equality axioms for $\overset{\circ}{=}$

The $(Gi)' \equiv (Gi)$ $(i=1,2,3)$ are provable in T by (T5). Since $(G4) \equiv u \overset{\circ}{=} v \wedge A(u) \rightarrow A(v)$ follows from this intuitionistically, so by I. $(G4)'$ has a solution in T too (which one also easily verifies in a direct way).

(6.2) For each formula A (of the functional language) provable in Heyting-arithmetic there is a functional φ in T with: $_T\vdash \varphi\text{-sol}(A')$.

Proof: We still have to complete the proof of (6.1) by (P1), (P2), (CI), (R1), (R2).

1. $(P1)' \equiv (P1) \equiv (T6)$; $(P2)' \equiv (P2)$ with (T7); $(R1)' \equiv (R1)$, $(R2)' \equiv (R2)$ by (5.2), (T4).

2. (CI)

Given are the φ-solution of $(A(0))' \equiv \bigvee x^{\alpha} \bigwedge y A_o(x,y,0,\underline{y})$ and the $\underline{\psi}$-solution of $(A(u^o) \rightarrow A(u'))' \equiv \bigvee x_1,x_2 \bigwedge y_1,y_2 \{A_o(y_1,x_2\underline{y},u,\underline{y}) \rightarrow A_o(x_1y_1,y_2,u',\underline{y})\}$, $u \notin \underline{y}$ in T:

(a) $A_o(\varphi\underline{y},w,0,\underline{y})$, $A_o(w_1,\psi_2u\underline{y}w_1w_2,u,\underline{y}) \rightarrow A_o(\psi_1u\underline{y}w_1,w_2,u',\underline{y})$

By (5.2), (T4): (b) $\chi 0\underline{y} \overset{\alpha}{=} \varphi\underline{y}$, $\chi u'\underline{y} = \psi_1 u^o \underline{y}(\chi u\underline{y})$

The substitution $w_1 : \chi u^o\underline{y}$, $w_2 : w$ in (a) yields with (b):

$$A_o(\chi 0\underline{y},w,0,\underline{y}),\quad A_o(\chi u^o\underline{y},\psi_2u\underline{y}(\chi u\underline{y})w,u,\underline{y}) \rightarrow A_o(\underbrace{\psi_1u\underline{y}(\chi u\underline{y})}_{=\chi u'\underline{y}},w,u',\underline{y})$$

For $tuw = \psi_2u\underline{y}(\chi u\underline{y})w$, $B(u,w) \equiv: A_o(\chi u\underline{y},w,u^o,\underline{y})$ thus:

$B(0,w)$, $B(u,tuw) \rightarrow B(u',w)$, $u \notin B(0,w),t$

From this in T by (T36): $B(u,w)$, i.e. $A_o(\chi u\underline{y},w,u^o,\underline{y})$. This is a χ-solution of $(A(u^o))' \equiv \bigvee x \bigwedge y A_o(x,y,u,\underline{y})$ in T.

(6.3) Classical arithmetic and number theory is functional interpretable in T.

Proof: (3.8),(6.2)

(6.4) The functional interpretation in the narrower sense of Heyting-analysis plus (MP), $(\underset{\rightarrow}{\bigvee})$ has a solution in T.

<u>Proof:</u> Additional to (6.1), (6.2) we have still to consider (ER)-qf, (AC), (MP) and $(\underset{\rightarrow}{\bigvee})$.

1. ((ER)-qf)' ≡ (ER)-qf with (T5)
2. $(AC)' \equiv (\bigwedge x \bigvee y \bigvee x_1 \bigwedge y_1 A_o(x_1,y_1,x,y,\underline{\underline{u}}) \rightarrow \bigvee z \bigwedge x \bigvee x_1 \bigwedge y_1 A_o(x_1,y_1,x,zx,\underline{\underline{u}}))'$

$\equiv (\bigvee z,x_2 \bigwedge x,y_1 A_o(x_2x,y_1,x,zx,\underline{\underline{u}}) \rightarrow \bigvee z,x_2 \bigwedge x,y_1 A_o(x_2x,y_1,x,zx,\underline{\underline{u}}))'$

$\equiv (B \rightarrow B)'$

By (6.1) this has a solution in T.

3. $(MP)' \equiv \bigvee x_1,..,x_6 \bigwedge y_1,..,y_6 \{((y_1(x_3\underline{\underline{y}})=0 \wedge A_o(y_2(x_3\underline{\underline{y}}),x_4\underline{\underline{y}},x_3\underline{\underline{y}},\underline{\underline{u}}))$

$\vee(y_1(x_3\underline{\underline{y}})\neq 0 \wedge \neg A_o(x_5\underline{\underline{y}},y_3(x_3\underline{\underline{y}})(x_5\underline{\underline{y}}),x_3\underline{\underline{y}},\underline{\underline{u}}))) \wedge$

$\neg A_o(x_6\underline{\underline{y}}(y_4(x_6\underline{\underline{y}})),y_5(x_6\underline{\underline{y}}),y_4(x_6\underline{\underline{y}}),\underline{\underline{u}}) \longrightarrow$

$\neg A_o(y_6,x_2\underline{\underline{y}},x_1y_1..y_5,\underline{\underline{u}})\}$

A $\underline{\varphi}$-solution for it is according to (5.2) performed in T by the following functionals:

$\varphi_1\underline{\underline{u}}v_1..v_5 = v_4v_2, \quad \varphi_2\underline{\underline{u}}\underline{\underline{v}} = v_3(v_4v_2)v_6$

$\varphi_3\underline{\underline{u}}\underline{\underline{v}} = v_4v_2, \quad \varphi_4\underline{\underline{u}}\underline{\underline{v}} = v_5v_2$

$\varphi_5\underline{\underline{u}}\underline{\underline{v}} = v_6, \quad \varphi_6\underline{\underline{u}}\underline{\underline{v}} = v_2$; because

$\{(v_1(\varphi_3\underline{\underline{u}}\underline{\underline{v}})=0 \wedge A_o(v_2(v_4v_2),v_5v_2,v_4v_2,\underline{\underline{u}})) \vee (v_1(\varphi_3\underline{\underline{u}}\underline{\underline{v}})\neq 0 \wedge$

$\neg A_o(v_6,v_3(v_4v_2)v_6,v_4v_2,\underline{\underline{u}}))\} \wedge \neg A_o(v_2(v_4v_2),v_5v_2,v_4v_2,\underline{\underline{u}}) \longrightarrow$

$\neg A_o(v_6,v_3(v_4v_2)v_6,v_4v_2,\underline{\underline{u}})$

4. $(\underset{\rightarrow}{\bigvee})' \equiv \bigvee x_1,..,x_9 \bigwedge y_1,..,y_9 \{((y_1(x_5\underline{\underline{y}})=0 \wedge A_o(y_2(x_5\underline{\underline{y}}),x_6\underline{\underline{y}},x_5\underline{\underline{y}},\underline{\underline{u}}_1)) \vee$

$(y_1(x_5\underline{\underline{y}})\neq 0 \wedge \neg A_o(x_7\underline{\underline{y}},y_3(x_5\underline{\underline{y}})(x_7\underline{\underline{y}}),x_5\underline{\underline{y}},\underline{\underline{u}}_1))) \wedge$

$(A_o(x_8\underline{\underline{y}}(y_6(x_8\underline{\underline{y}})(x_9\underline{\underline{y}})),y_7(x_8\underline{\underline{y}})(x_9\underline{\underline{y}}),y_6(x_8\underline{\underline{y}})(x_9\underline{\underline{y}}),\underline{\underline{u}}_1) \rightarrow$

$B_o(y_5(x_8\underline{\underline{y}}),x_9\underline{\underline{y}},y_4(x_8\underline{\underline{y}}),\underline{\underline{u}}_2)) \rightarrow A_o(y_8(x_3\underline{\underline{y}}),x_4\underline{\underline{y}},x_3\underline{\underline{y}},\underline{\underline{u}}_1) \rightarrow$

$B_o(x_2y_1..y_8,y_9,x_1y_1..y_7,\underline{\underline{u}}_2)\}$

The following functionals according to (5.2) perform a $\underline{\varphi}$-solution for this in T:

$\varphi_1\underline{\underline{u}}v_1..v_7 = v_4v_2, \quad \varphi_2\underline{\underline{u}}v_1..v_8 = v_5v_2$

$\varphi_3\underline{\underline{u}}\underline{\underline{v}} = v_6v_2v_9, \quad \varphi_4\underline{\underline{u}}\underline{\underline{v}} = v_3(v_6v_2v_9)(v_8(v_6v_2v_9))$

$\varphi_5\underline{\underline{u}}\underline{\underline{v}} = v_6v_2v_9, \quad \varphi_6\underline{\underline{u}}\underline{\underline{v}} = v_7v_2v_9$

$\varphi_7\underline{\underline{u}}\,\underline{\underline{v}} = v_8(v_6v_2v_9)$, $\varphi_8\underline{\underline{u}}\,\underline{\underline{v}} = v_2$, $\varphi_9\underline{\underline{u}}\,\underline{\underline{v}} = v_9$; because

$$\{(v_1(\varphi_5\underline{\underline{u}}\,\underline{\underline{v}})=0 \wedge A_o(v_2(v_6v_2v_9), v_7v_2v_9, v_6v_2v_9, \underline{\underline{u}}_1)) \vee (v_1(\varphi_5\underline{\underline{u}}\,\underline{\underline{v}}) \neq 0 \wedge$$
$$\neg A_o(v_8(v_6v_2v_9), v_3(v_6v_2v_9)(v_8(v_6v_2v_9)), v_6v_2v_9, \underline{\underline{u}}_1))\} \wedge$$
$$\{A_o(v_2(v_6v_2v_9), v_7v_2v_9, v_6v_2v_9, \underline{\underline{u}}_1) \rightarrow B_o(v_5v_2, v_9, v_4v_2, \underline{\underline{u}}_2)\} \rightarrow$$
$$A_o(v_8(v_6v_2v_9), v_3(v_6v_2v_9)(v_8(v_6v_2v_9)), v_6v_2v_9, \underline{\underline{u}}_1) \rightarrow B_o(v_5v_2, v_9, v_4v_2, \underline{\underline{u}}_2)$$

From (6.4) follows with (3.9), (3.10):

(6.5) Classical arithmetic plus (ER)-qf, (AC)-qf is functional interpretable in T. Because of (3.13), (1.2) already here (4.5) is valid.

In surveying the proofs of this chapter one establishes that the functional interpretation depends only at one point on the prime formulae, namely for (Taut)-$\wedge \equiv (A \rightarrow A \wedge A)$. As by (4.6) there are other provable formulae whose functional interpretation always depends on the prime formulae, so these formulae can only be proved in the present codification even in Heyting-analysis plus (MP), ($\underset{\rightarrow}{V}$) by the help of (Taut)-$\wedge$.

We still mention the intensional treatment of the so far given functional interpretation. As noticed above, we will not enter into this matter; but the reader may convince himself that the given arguments, with the exception of complete induction (CI), everywhere carry over to a treatment in tupels. An analogous intensional treatment of (CI) requires for (b) in the proof of (6.2) and for the $\underline{\underline{\psi}}$-definition in the proof of (T36) the simultaneous recursion

$$\varphi_i 0 = \psi_i$$
$$\varphi_i u' = \chi_i(\varphi_1 u)\ldots(\varphi_n u)u^o \qquad i=1,..,n \quad \text{for preceding } \psi_1, \chi_1, \ldots, \psi_n, \chi_n$$

By (T42) this simultaneous recursion is in T intensional derivable. Altogether this yields in T a full intensional treatment of the functional interpretation for classical arithmetic over all finite types plus (AC)-qf as well as of the functional interpretation in the narrower sense for Heyting-analysis over all finite types plus (MP), ($\underset{\rightarrow}{V}$) - in each case of course without (ER)-qf.

Concluding yet a word on the proof-theoretic strength s of T. Since by T also number theory is consistent so $s \geqslant \varepsilon_o$. The consistency of T follows from the extensional computability of all closed type 0 terms of T in a manner which can be formalized in number theory (see chapter X). But the computability of the closed type 0 terms of T can be proved in number theory plus transfinite induction up to ε_o according to Kreisel [26], Tait [52], Hinata [17], Schütte [48]. Thus T, classical number theory with or without (ER)-qf, (AC)-qf, the intuitionistic number theory and all intermediate systems up to Heyting-analysis plus (ER)-qf, (MP), ($\underrightarrow{V}$) have the same proof-theoretic strength, namely ε_o.

VII. The calculus T∪BR of the bar recursive functionals

Because T is explicitly closed against course-of-values recursions and manyfold nested recursions on number variables (Diller [5], 17-21) for genuine extensions of T new principles are necessary. Gödel proposed already in [12], p. 286 the enlargement by transfinite types or by the method of reasoning Brouwer used in the proof of the bar resp. fan theorem.

An expansion of functional interpretation to analysis raises the question to the nature of the functional domain needed. The results of Kreisel [23] are:

1.(p. 120f) Over a functional domain B, which contains T and is closed against number choice sequences and (AC)-qf, classically holds for formulae A of second order: $A \leftrightarrow \bar{A} \leftrightarrow (\bar{A})^{!}_{B}$. Here C_B denotes the quantifier relativization of C to B.

2. Classically in the model $\mathfrak{M}_K$ of continuous functionals (Kreisel [23], 114-116, 125-127; Kleene [20]) - which reflects T - $((AC)\text{-qf})_{\mathfrak{M}_K}$ is fulfilled (indeed the choice is recursively continuous).

Altogether for an analysis formulated in the functional language of second order: classically $A \leftrightarrow (\bar{A})^{!}_{\mathfrak{M}_K}$.

Kreisel conjectured (see [27], 145-146; [28], 348) that analogous to number theory classical logic is eliminable herein for <u>provable</u> A so that

classical analysis $\vdash A \implies (\bar{A})^{!}_{\mathfrak{M}_K}$ intuitionistic

where $\mathfrak{M}_K$ taken intuitionistically contains besides constants also freely chosen neighborhood functionals.

A fundamental step in making these intentions concrete was done by Spector 1961 in his last paper [50]. Using the continuity of constructive and intuitionistic functionals for well-foundedness ($t(\varphi^{\alpha o})^{o}$ continuous in $\varphi \iff t(\varphi)$-value depends only on a finite initial segment of φ, i.e. $\bigwedge f^{\alpha o} \bigvee x^{o} \bigwedge g^{\alpha o}\ t(\overline{f,x}) \overset{o}{=} t(\overline{f,x * g})$) he formulated bar recursion as the recursion principle corresponding to Brouwer's bar induction over the natural numbers (first formalized by Kleene [22] §6) but now generalized to all finite types

$$(*) \quad \phi w_1 w_2 w_3 u^o v^{\alpha o} = \begin{cases} w_2 u(\overline{v,u}) & \text{if } w_1(\overline{v,u})<u \\ w_3 u(\overline{v,u})(\lambda x^{\alpha} \phi w_1 w_2 w_3 u'(\overline{v,u*x})) & \text{otherwise} \end{cases}$$

and he showed that a functional interpretation of classical $(AC)^o$-analysis (over the functional language of all finite types) can be given in $T\cup(*)$.

Over T $(*)$ is equivalent (see below (TB1)) to

$$(BR)^{\alpha} \begin{cases} w_1(\overline{v,u})<u \rightarrow \phi w_1 w_2 w_3 u^o v^{\alpha o} = w_2 \\ w_1(\overline{v,u})\geq u \rightarrow \phi w_1 w_2 w_3 u \; v \;\; = w_3(\lambda x^{\alpha} \phi w_1 w_2 w_3 u'(\overline{v,u*x})) \end{cases}$$

That under certain conditions the functional ϕ is attainable determined by these relations, can be proved with the help of $(BI)^{\alpha}_{D}$ as follows:

Obviously it suffices to show this for $\phi w_1 w_2 w_3 u(\overline{v,u})$.

$A(v_1^{\alpha o};u_1^o) \equiv: w_1(<>v(\lambda x^o v_1(-ux))u)<+uu_1$

$B(v_1^{\alpha o};u_1^o) \equiv: \phi w_1 w_2 w_3(+uu_1)(<>v(\lambda x^o v_1(-ux))u)$ defined

$\bigwedge x^{\alpha o} \bigvee y^o A(\overline{x,y};y)$ is ensured in any case by the use of continuous functionals; nevertheless well-foundedness may already occur previously depending on the course of w_1-values.

$A(v_1;u_1) \vee \neg A(v_1;u_1)$ is valid by (T21) if computable functionals are used.

$A(\overline{v_1,u_1};u_1) \rightarrow B(\overline{v_1,u_1};u_1)$ follows extensionally from the first line of $(BR)^{\alpha}$.

$\bigwedge x^{\alpha} B(\overline{v_1,u_1*x};u_1') \rightarrow B(\overline{v_1,u_1};u_1)$ is proved extensionally from the second line of $(BR)^{\alpha}$ provided that the first line does not already apply.

From this by $(BI)^{\alpha}_{D}$ one gets $B(\sigma;0)$, i.e. $\phi w_1 w_2 w_3 u(\overline{v,u})$ is defined.

We consider the (BR)-equations not only as a general specification of the functional ϕ (namely as the unique solution of (BR)) but still stronger as a constructive computability instruction for ϕ. - The above used $(BI)^{\alpha}_{D}$-application does not refer to arbitrary species but only to the above employed properties of computable continuous functionals. As against Brouwer's bar induction however now all finite types are concerned.

The problem remains to find concretely a continuous functional domain where the T∪BR-equations are valid, which (on account of the in advance incalculable possibilities of BR-well-foundedness) is closed against choice sequences in arbitrary types and for which the needed $(BI)_D$-applications can be established with more insight.

Howard [18] extended Spector's interpretation result to (ωAC)-analysis. In analogy to (CI) he gives a solution of the functional interpretation in the narrower sense of $(BI)_M^\alpha$ in $T\cup(BR)^\alpha$. Because (ωAC) reduces classically with (ER)-qf, (AC)-qf by raising types to (ωAC)-Λ and because by Howard-Kreisel [19], 352 classically $(BI)_M^\alpha\text{-}V \vdash (\omega AC)^\alpha\text{-}\Lambda$ this altogether gives - on account of $(BI)_M\text{-}V$, $(MP) \vdash \neg\neg((BI)_M\text{-}V)^*$ intuitionistically - a solution of the functional interpretation in the narrower sense of $(\neg\neg(\omega AC)^*)'$ in $T\cup(BR)$. - In this paper Howard moreover shows that these interpretation results also follow with the $(BI)_M$-rule and the so-called (BR)-rule (i.e.(BR) for <u>fixed</u> preceding w_1, w_2, w_3) since $((BI)_M\text{-rule})^{\alpha(\alpha o)}\ {}_T\!\vdash (BI)_M^\alpha$, $((BR)\text{-rule})^{\alpha(\alpha o)}\ {}_T\!\vdash (BR)^\alpha$.

A constructive proof of computability was so far only given for $T\cup(BR)^o$, namely in

Tait [54] with the help of "step notions"

Diller [4], [5] for $(BR)^o$-rule together with an ordinal estimate

Kreisel [27], 151 on the basis of generalized inductive definition .

These proofs can be formalized in today known intuitionistic analysis. In this deductive framework one still can use the means of Howard-Kreisel [19] §7 to infer the computability of $T\cup BR^{o\cdots o}$ from a suitable intuitionistic model. However higher (BR)-cases cannot be established in this way because so far known and formalized intuitionistic analysis is provably consistent in classical Π_1^1-analysis (Kreisel [27], 151) and a functional interpretation of Π_1^1-comprehension can be given with $(BR)^{o(oo)}$. Thus foundations for bar recursion of higher type require stronger principles than so far accepted.

Kreisel stated in [27], 147 without proof that the continuous functionals are classically a model of full T∪BR. An analogous assertion was proved rigorously by Scarpellini [38] for a better manageable classical topological model of T∪BR. Meanwhile also Tait [56] published a classical treatment of T∪BR.

Thus higher bar recursion is at least classically satisfiable. The problem remains in how far this can also be realized constructively. A set-up - starting from a conjecture of Gödel in 1962 - is described in Tait [55]. It is a matter of generalizing the formation operation for the intuitionistic continuous functionals in the spread over the natural numbers to spreads over species. By introducing such inductive defined "higher" species (namely type α-valued functionals for choice sequences over type β-functionals) into intuitionistic mathematics now the problem is to build up a model for T∪BR that on the one side is not "too large" to be still constructive and on the other side is provably closed against T∪BR. Tait reports in [55],197 that in generalizing the hitherto known methods in this sense he was able to prove the computability of $T{\cup}BR^{\theta_i}$, where $\theta_o{\equiv}:0$, $\theta_{m+1}{\equiv}:0(\theta_m 0)$ (because of $\theta_1{\equiv}0(00)$ this covers Π^1_1-analysis). However proofs are not published till now.

In the following we try to describe the main points of the so far obtained results. We don't enter into the model $\mathfrak{M}_K$ of the continuous functionals, because we will always keep the direct functional structure. Also we don't take Howard's functional interpretation but extend Spector's original proceeding from $(AC)^o$-analysis to (ωAC)-analysis because in our opinion this is likewise perspicuous. Computability is shown with the help of "step notions" at first for $T{\cup}BR^o$ and then after Howard-Kreisel for $T{\cup}BR^{o\cdots o}$. It follows a discussion of generalizations of the methods so far employed. Finally in the last chapter a constructive model is given which covers all types. - Scarpellini [40] after having seen our work was able to improve his topological model in a constructive way thus yielding another proof by the same means (see chapter XIV) plus an additional axiom of choice; but it is fair to say that our model gives a direct computational analysis of the situation and is therefore more minimal (indeed it is minimal with respect to the methods used).

The calculus T∪BR results from T by adding to the language the functional signs ϕ_o, ϕ_1, ... which below are associated in a unique manner to the defining equations (BR) of bar recursion. It is not possible to put the (BR)-scheme in the above given form into the deduction frame, because in our calculi the λ-operator is only implicit at hand and (BR) above uses λ essentially explicit. However this can be achieved by remodeling the places to which this λ-operator refers as explicit new arguments. So now to the old $\phi w_1 w_2 w_3 u^o\langle vo,..,v(-2u),w,vu,...\rangle$, $\phi w_1 w_2 w_3 u^o v^{\alpha o}$ correspond the new $\phi w_1 w_2 w_3 u^o v^{\alpha o} w^\alpha$, $\phi w_1 w_2 w_3 uv(v(-1u))$. Therefore we add to the deduction frame of T the following axiom scheme $(BR)^{\alpha}$:

$$(BR)^{\alpha}\begin{cases} w_1(\overline{\overline{v,-1u}*w,u})<u \longrightarrow \phi w_1^{o(\alpha o)} w_2 w_3 u^o v^{\alpha o} w^{\alpha} = w_2 \\ w_1(\overline{\overline{v,-1u}*w,u})\geq u \longrightarrow \phi w_1 w_2 w_3 uvw = w_3(\phi w_1 w_2 w_3 u'(\overline{\overline{v,-1u}*w,u})) \end{cases}$$

It seems that so far nobody has observed that a precise formalization of all details in Spector's proof [50] §10 using $\overset{o}{=}$ only requires an additional <u>constructive ω-rule.</u> The following concrete version suffices.

(ω) If it can be shown by means of intuitionistic arithmetic that $A(z)$ is provable for <u>all</u> numerals z then also $A(u^o)$ (variable u) is provable.

(ω) is used only in one place in the proof of (TB6). This proof can likewise be given without (ω) if higher definitional equalities $\overset{\alpha}{=}$ with the usual equality relations are introduced. But since the reasoning needs certain extensionalities moreover in that case we have to require

$$\frac{A \longrightarrow r\underline{u} \overset{\beta}{=} s\underline{u}}{A \longrightarrow r \overset{\alpha}{=} s} \quad (\underline{u} \notin A,r,s) \text{ for certain } \alpha,\beta.$$

The computability proof of this calculus now has to employ considerations from Tait [54]. This is avoided here by using only $\overset{o}{=}$ and by giving the interpretation from the usual extensional standpoint. - Obviously (ω) covers (CI).

Let $T \cup BR$ be this extended calculus. All results of chapter V carry over because only explicit proofs have been given. - As before the notation refers to <u>all</u> types which make sense to a context.

$(TB1)^{\alpha}$ In $T \cup BR$ there are functionals ϕ with:

$$\begin{cases} w_1(\overline{\overline{v,-1u}*w,u})<u \longrightarrow \phi w_1 w_2 w_3 u^o v^{\alpha o} w^{\alpha} = w_2 uvw \\ w_1(\overline{\overline{v,-1u}*w,u})\geq u \longrightarrow \phi w_1 w_2 w_3 uvw = w_3 uvw(\phi w_1 w_2 w_3 u'(\overline{\overline{v,-1u}*w,u})) \end{cases}$$

<u>Proof:</u> Let us denote temporarily the wanted ϕ by ϕ_1 and the ϕ from (BR) with ϕ_o.

(1) $\psi_1 v_1 u^o v^{\alpha o} w^{\alpha} u_1^{\alpha} =: v_1 u_1 u'(\overline{\overline{v,-1u}*w,u}) u_1$

(2) $\psi_2 w_3 v_1 u^o v^{\alpha o} w^{\alpha} =: w_3 uvw(\psi_1 v_1 uvw)$

(3) $\phi_1 w_1 w_2 w_3 u^o v^{\alpha o} w^{\alpha} =: \phi_o w_1 w_2(\psi_2 w_3) uvwuvw$

With this one gets from (BR) by (S):

$$w_1(\overline{\overline{v,-1u}*w,u})<u \longrightarrow \phi_1 w_1 w_2 w_3 uvw = \phi_o w_1 w_2(\psi_2 w_3)uvwuvw = w_2 uvw \qquad (3),(BR)$$

$$w_1(\overline{\overline{v,-1u}*w,u}) \geqslant u \longrightarrow \phi_1 w_1 w_2 w_3 uvw = \psi_2 w_3(\phi_o w_1 w_2(\psi_2 w_3)u'(\overline{\overline{v,-1u}*w,u}))uvw$$

$$= w_3 uvw(\psi_1(\phi_o w_1 w_2(\psi_2 w_3)u'(\overline{\overline{v,-1u}*w,u}))uvw)$$

$$= w_3 uvw(\phi_1 w_1 w_2 w_3 u'(\overline{\overline{v,-1u}*w,u})) \qquad (3),(BR),(2),(4),(E)$$

since for a variable u_1 holds:

$$(4) \quad \psi_1(\phi_o w_1 w_2(\psi_2 w_3)u'(\overline{\overline{v,-1u}*w,u}))uvwu_1$$

$$= \phi_o w_1 w_2(\psi_2 w_3)u'(\overline{\overline{v,-1u}*w,u})u_1 u'(\overline{\overline{v,-1u}*w,u})u_1 \qquad (1)$$

$$= \phi_1 w_1 w_2 w_3 u'(\overline{\overline{v,-1u}*w,u})u_1 \qquad (3)$$

<u>Remark:</u> Conversely also (BR) comes from (TB1), namely $\phi_o w_1 w_2 w_3 = \phi_1 w_1(\varphi_1 w_2)(\varphi_2 w_3)$ with $\varphi_1 w_2 uvw = w_2$, $\varphi_2 w_3 uvw = w_3$.

The functional interpretation of 'analysis will be settled by using only the following special case of (TB1). <u>These BR-functionals $(TB2)^\alpha$ are in the following always denoted by $\phi^{(\alpha)}$.</u>

$(TB2)^\alpha$ In T∪BR there are functionals ϕ with:

$$w_1(\overline{\overline{v,-1u}*w,u})<u \longrightarrow \phi w_1^{o(\alpha o)} w_2 u^o v^{\alpha o} w^\alpha \overset{\alpha o}{=} \overline{\overline{v,-1u}*w,u}$$

$$w_1(\overline{\overline{v,-1u}*w,u}) \geqslant u \longrightarrow \phi w_1 w_2^{\alpha(\alpha o\alpha)o} uvw = \langle\rangle(\overline{\overline{v,-1u}*w,u})(\phi w_1 w_2 u'(\overline{\overline{v,-1u}*w,u})\, a_o(u,v,w))u$$

$$\text{with } a_o(u,v,w) \equiv: w_2 u(\phi w_1 w_2 u'(\overline{\overline{v,-1u}*w,u}))$$

<u>Proof:</u> If ϕ_1 denotes the ϕ of $(TB1)^\alpha$, so one gets (TB2) as follows: $\phi w_1 w_2 =: \phi_1 w_1 \psi_1(\psi_2 w_2)$, where $\psi_1 uvw =: \overline{\overline{v,-1u}*w,u}$; $\psi_2 w_2 uvwv_1 =: \langle\rangle(\overline{\overline{v,-1u}*w,u})(v_1(w_2 uv_1))u$

Now let us get an <u>informal</u> survey over those properties of bar recursion with which we will be concerned below. At this w_1, w_2 are considered as fixed; ϕ is short for $\phi w_1 w_2$. Further take $\langle\ldots\rangle_u \equiv: \sigma$ if $u<0$, $\langle\ldots\rangle_u \equiv: \langle\ldots\rangle$ if $u \geqslant 0$; let $*$ arrange the prefixing of a finite sequence. - In this notation (TB2) can be writen informally as

$$(1) \quad \begin{cases} w_1(\langle v0,..,v(u-2),w\rangle_{u-1})<u \longrightarrow \phi uvw = \langle v0,..,v(u-2),w\rangle_{u-1} \\ w_1(\langle v0,..,v(u-2),w\rangle_{u-1}) \geqslant u \longrightarrow \phi uvw = \langle v0,..,v(u-2),w\rangle_{u-1} * \\ \qquad \langle \phi u'\langle v0,..,v(u-2),w\rangle_{u-1} a_o(u,v,w)\overset{+}{u}, \ldots\rangle \end{cases}$$

Here $\downarrow$ points to the solely varying index.

From that one has directly

(2) $z<-1u \longrightarrow \phi uvwz=vz, \quad u\neq 0 \wedge z=-1u \longrightarrow \phi uvwz=w$

To register the regressive instructions in (1) we continue the a_o-definition recursively as follows:

$$(3)\begin{cases} a_o(u,v,w) = w_2u(\phi u'\langle v0,\ldots,v(u-2),w\rangle_{u-1}) \\ a_{i+1}(u,v,w) \equiv: w_2(u+i+1)(\phi(u+i+1)'(\langle v0,\ldots,v(u-2),w\rangle_{u-1} * \\ \qquad \langle a_o(u,v,w),\ldots,a_i(u,v,w)\rangle)) \end{cases}$$

Induction on i yields

(4) $a_i(u,v,w) = a_i(u,\langle v0,\ldots,v(u-2)\rangle,w), \quad a_i(0,\sigma,\sigma) = a_i(0,v,w)$

Because of (2) <u>over a continuous functional domain</u> now we have the following retrograde computation according to (1):

$$(5)\begin{cases} \phi uvw = \langle v0,\ldots,v(u-2),w\rangle_{u-1} * \langle a_o(u,v,w),\ldots,a_{z-1}(u,v,w)\rangle \\ \qquad = \phi u(\langle v0,\ldots,v(u-2)\rangle)w & (4) \\ \text{with } z = \mu i(w_1(\langle v0,\ldots,v(u-2),w\rangle_{u-1} * \langle a_o(u,v,w),\ldots,a_{i-1}(u,v,w)\rangle) < u+i) \end{cases}$$

We are especially interested in bar recursion beginning with 0:

$\phi_o =: \phi 0\sigma\sigma; \quad b_i =: a_i(0,\sigma,\sigma), \quad b_{-i} =: \sigma$

$$(6)\begin{cases} \phi_o = \phi 0\sigma\sigma = \langle b_o,\ldots,b_{z_o-1}\rangle = \phi 0vw & (5),\ (4) \\ \text{with } z_o = \mu i(w_1(\langle b_o,\ldots,b_{i-1}\rangle)<i) \end{cases}$$

The a_i, b_i are related in the following way:

(7) $\underbrace{a_i(u,\langle b_o,\ldots,b_{u-2}\rangle,b_{u-1})}_{c_i^u \equiv:} = b_{i+u}$

This is proved by induction on i:

I. $a_o(u,\langle b_o,\ldots,b_{u-2}\rangle,b_{u-1}) = w_2u(\phi u'\langle b_o,\ldots,b_{u-1}\rangle) = b_u$ (3), what follows by induction on u with (3).

II. $a_{i+1}(u,\langle b_o,\ldots,b_{u-2}\rangle,b_{u-1}) = w_2(u+i+1)(\phi(u+i+1)'(\langle b_o,\ldots,b_{u-1}\rangle *$
$\langle a_o(u,\langle b_o,\ldots,b_{u-2}\rangle,b_{u-1}),\ldots,a_i(u,\langle b_o,\ldots,b_{u-2}\rangle,b_{u-1})\rangle))$
$=w_2(u+i+1)(\phi(u+i+1)'(\langle b_o,\ldots,b_{u-1},b_u,\ldots,b_{i+u}\rangle))$

$=b_{i+u+1}$ (3),ind.hyp.

From this now

$$u<z_o \rightarrow \phi u(<\phi_o 0,\dots,\phi_o(u-2)>)(<\phi_o(u-1)>_{u-1}0) = \quad (6)$$

$$\phi u(<b_o,\dots,b_{u-2}>)(<b_{u-1}>_{u-1}0) = <b_o,\dots,b_{u-1}>*<c^u_o,\dots,c^u_{z-1}> \quad (5)$$

$$\text{for } z = \mu i(w_1(<b_o,\dots,b_{u-1}>*<c^u_o,\dots,c^u_{i-1}>)<u+i)$$

$$= <b_o,\dots,b_{u-1},b_u,\dots,b_{z_o-1}> = \phi_o \quad (7),(6)$$

$$z_o<u \rightarrow \phi u(<\phi_o 0,\dots,\phi_o(u-2)>)(<\phi_o(u-1)>_{u-1}0) = \phi u\phi_o\sigma = \phi_o \quad (6),(1)$$

Thus: (8) $\phi u(<\phi_o 0,\dots,\phi_o(u-2)>)(<\phi_o(u-1)>_{u-1}0) = \phi_o$

For $n_o =: w_1\phi_o = w_1(<b_o,\dots,b_{z_o-1}>)<z_o$ (6); because of the minimality in (6) (9) $w_1(<b_o,\dots,b_{n_o-1}>)\geq n_o$. From this ensues in particular

$$(10)\ \phi_o n_o = b_{n_o} = a_o(n_o,<b_o,\dots,b_{n_o-2}>,b_{n_o-1}) = w_2 n_o(\phi n_o'<\phi_o 0,\dots,\phi_o(n_o-1)>)$$

(6),(7),(3)

In chapter VIII it is shown that the relations (2), (8), (10) suffice for a functional interpretation of analysis. Therefore it is now the point to reflect these considerations in the calculus T∪BR. The above argumentation essentially comes off in the extension of T by (1) and the used μ-applications. But these μ-cases can be derived in T∪BR from (TB1):

$$\Gamma w_1 u^o v^{\alpha o} w^\alpha \ o(\underset{=}{\alpha}\underset{.}{o}) \begin{cases} \sigma & \text{if } w_1(<v0,\dots,v(u-2),w>_{u-1})<u \\ \lambda x^{\alpha o}(1+\Gamma w_1 u'<v0,\dots,v(u-2),w>_{u-1}(xu)) & \text{otherwise} \end{cases}$$

$$\mu i(w_1(<v0,\dots,v(i-1)>)<i) =: \Gamma w_1 0\sigma\sigma v$$

We still remark that the above considerations with the exception of the last equality in (10) are also valid for arbitrary

$$a_i(u,v,w) = f(u+i,<v0,\dots,v(u-2),w>_{u-1}*<a_o(u,v,w),\dots,a_{i-1}(u,v,w)>)$$

Thus (BR) is not fully used in that part; for the infinite regress within the spread which is characteristic of bar recursion is here without effect since also the used μ-applications rest on a very weak (BR) by regressing only on a single path through the underlying spread. Not before chapter VIII (BR) is fully used by means of the given a_o-definition in the form of the last equality in (10).

The now following derivations in T∪BR of the relations needed are not given in the above described manner but on a shorter and already by Spector gone way which needs no μ-operator. The employed (E)-applications refer in their really extensional part only to the equality of ω-sequences over an arbitrary type provided all corresponding members are equal ("sequence-(E)").

(TB3) $z^{o}<-1u \longrightarrow \phi w_1 w_2 uvwz \overset{\alpha}{=} vz, \quad u\neq 0 \wedge z \overset{o}{=} -1u \longrightarrow \phi w_1 w_2 uvwz \overset{\alpha}{=} w$

Proof:

$u\neq 0 \longrightarrow u=(\delta u)'<u' \wedge \delta u<u \wedge -1u=-0'(\delta u)'=\delta u$ (T8),(T16),(T14)

$(*) \longrightarrow -1u=\delta u<u$

$z<-1u \longrightarrow -1u\neq 0 \wedge 1<u \wedge u\neq 0 \wedge -1u<u \wedge z<u$ (T37),(*),(T31)

$\longrightarrow \phi w_1 w_2 uvwz=\overline{\overline{v,-1u*w},u}z=(\overline{v,-1u*w})z=vz$ (TB2),(T13),(T39),(T40),(T41)

$u\neq 0 \wedge z=-1u \longrightarrow z<u$ (*)

$\longrightarrow \phi w_1 w_2 uvwz=\overline{\overline{v,-1u*w},u}z=(\overline{v,-1u*w})z=w$ (TB2),(T13),(T39),(T40),(T41)

(TB4) $\overline{\overline{r,u},u^{o}} = \overline{r,u}, \quad \overline{r,0} = \sigma$

Proof:

$z<u \longrightarrow \overline{\overline{r,u},u}z=\overline{r,u}z=rz, \quad z\geqslant u \longrightarrow \overline{\overline{r,u},u}z=\sigma=\overline{r,u}z, \quad \overline{r,0}z=\sigma z$ (T40)

Sequence-(E) now gives the assertion.

(TB5) If $u=0 \vee s=r(-1u)$, then $\overline{\overline{r,-1u*s},u} = \overline{r,u}$

Proof:

$z<u \longrightarrow u\neq 0 \wedge z<u=(\delta u)' \wedge z\leqslant \delta u=-1u$ (T37), (T8),(T38),(*) in (TB3)

$\longrightarrow z<u \wedge (z<-1u \vee z=-1u) \wedge u\neq 0$ (T29)

$\longrightarrow \overline{\overline{r,-1u*s},u}z=(\overline{r,-1u*s})z=rz \vee \overline{\overline{r,-1u*s},u}z=(\overline{r,-1u*s})z=s=r(-1u)=rz$ premise,(T40),(T41)

$\longrightarrow \overline{\overline{r,-1u*s},u}z=rz=\overline{r,u}z$ (T40)

$z\geqslant u \longrightarrow \overline{\overline{r,-1u*s},u}z=\sigma=\overline{r,u}z$ (T40)

(T13), sequence-(E) now yield the conclusion of (TB5).

$(\phi_o)^{\alpha} \quad \phi_o w_1 w_2 \overset{\alpha o}{=} \phi w_1 w_2 0\sigma\sigma$

$(\nu_o)^\alpha \quad \nu_o w_1 w_2 \overset{o}{=} w_1(\phi_o w_1 w_2)$

$(TB6)^\alpha \; \phi_o w_1 w_2 = \phi w_1 w_2 u^o(\overline{\phi_o w_1 w_2,-1u})(\overline{\lambda x^o \phi_o w_1 w_2(-1u),u0})$

<u>Proof:</u> We show by metainduction that the assertion is valid for every numeral z instead of u^o. (TB6) follows from this by (ω).

I. $\phi w_1 w_2 0(\overline{\phi_o w_1 w_2,-10})(\overline{\lambda x^o \phi_o w_1 w_2(-10),00}) = \phi w_1 w_2 0\overline{0}\overline{0} = \phi_o w_1 w_2$ (TB4), because by (T37) $-u0=0$.

II. Under the abbreviations

$r \overset{\alpha o}{\equiv}: \overline{\phi_o w_1 w_2,-1z}, \quad s \overset{\alpha}{\equiv}: \overline{\lambda x^o \phi_o w_1 w_2(-1z),z0}$

$r_1 \overset{\alpha o}{\equiv}: \overline{\phi_o w_1 w_2,-1z'}, \quad s_1 \overset{\alpha}{\equiv}: \overline{\lambda x^o \phi_o w_1 w_2(-1z'),z'0}$

$V \; \equiv: (\phi w_1 w_2 z r s = \phi_o w_1 w_2), \quad B \equiv: (\phi w_1 w_2 z' r_1 s_1 = \phi_o w_1 w_2)$

we have to show: V provable $\Longrightarrow$ B provable.

(1) $z \neq 0 \rightarrow z = (\delta z)' > 0 \wedge s = \phi_o w_1 w_2(-1z)$ (T8),(T37),(T40)

(2) $s_1 = \phi_o w_1 w_2(-1z') = \phi_o w_1 w_2 z$ (T37),(T40),(T14)

(3) $\overline{\overline{r,-1z}*s,z} = \overline{\overline{\phi_o w_1 w_2,-1z}*s,z} = \overline{\phi_o w_1 w_2,z} = \overline{\phi_o w_1 w_2,-1z'} = r_1$ (TB4), (1),(TB5),(T21),(T14)

(4) $\overline{\overline{r_1,-1z'}*s_1,z'} = \overline{\overline{\phi_o w_1 w_2,-1z'}*s_1,z'} = \overline{\phi_o w_1 w_2,z'}$ (TB4),(2),(TB5)

$w_1(\overline{\overline{r,-1z}*s,z}) < z \rightarrow \phi_o w_1 w_2 = \phi w_1 w_2 z r s = \overline{\overline{r,-1z}*s,z} = \overline{\phi_o w_1 w_2,z}$ premise V, (TB2), (3)

$\rightarrow \overline{\phi_o w_1 w_2,z}u^o = \phi_o w_1 w_2 u = \overline{\phi_o w_1 w_2,z'}u$ (T40),(T13), (T16),(T31),(T29)

$\rightarrow \overline{\overline{r,-1z}*s,z} = \overline{\phi_o w_1 w_2,z} = \phi_o w_1 w_2 = \overline{\phi_o w_1 w_2,z'}$

$= \overline{\overline{r_1,-1z'}*s_1,z'}$ (3),sequence-(E),(4)

$\rightarrow w_1(\overline{\overline{r_1,-1z'}*s_1,z'}) < z < z'$ (T31),(T16)

$\rightarrow \phi w_1 w_2 z' r_1 s_1 = \overline{\overline{r_1,-1z'}*s_1,z'} = \phi_o w_1 w_2$ (TB2)

(5) $\rightarrow B$

$w_1(\overline{\overline{r,-1z*s,z}})\geqslant z \longrightarrow \phi_o w_1 w_2 = \phi w_1 w_2 zrs = <>(\overline{\overline{r,-1z*s,z}})(\phi w_1 w_2 z'(\overline{\overline{r,-1z*s,z}})$

$a_o)z$

premise V,(TB2)

$\longrightarrow (u^o<z=-1z' \longrightarrow \phi_o w_1 w_2 u = \overline{\overline{r,-1z*s,zu}} =$

$\phi w_1 w_2 z'(\overline{\overline{r,-1z*s,z}})a_o u)$ (T14),(T39),(TB3)

$\wedge(u^o\geqslant z \longrightarrow \phi_o w_1 w_2 u = \phi w_1 w_2 z'(\overline{\overline{r,-1z*s,z}})a_o u)$ (T39)

$\wedge\phi_o w_1 w_2 z = \phi w_1 w_2 z'(\overline{\overline{r,-1z*s,z}})a_o z = a_o$ (T39),(T28),

(TB3),(T6),(T14)

$\longrightarrow \phi_o w_1 w_2 = \phi w_1 w_2 z'(\overline{\overline{r,-1z*s,z}})(\phi_o w_1 w_2 z) = \phi w_1 w_2 z' r_1 s_1$

(T13),sequence-(E),(3),(2)

(6) $\longrightarrow$ B

From (5),(6) follows with (T13): B.

$(TB7)^\alpha$ $\phi w_1 w_2 u'(\overline{\phi_o w_1 w_2,u})(\phi_o w_1 w_2 u) = \phi_o w_1 w_2$

<u>Proof:</u> (TB6),(T14),(T37),(T4o)

$(TB8)^\alpha$ $0\leqslant u\leqslant \nu_o w_1 w_2 \longrightarrow w_1(\overline{\overline{\phi_o w_1 w_2,-1u*(\lambda x^o \phi_o w_1 w_2(-1u),u0)}},u)\geqslant u$

<u>Proof:</u>

$s \overset{\alpha}{\equiv}: \overline{\lambda x^o \phi_o w_1 w_2(-1u),u0}$

$0\leqslant u\leqslant \nu_o w_1 w_2 \wedge w_1(\overline{\overline{\phi_o w_1 w_2,-1u*s,u}})<u$

$\longrightarrow \phi_o w_1 w_2 = \phi w_1 w_2 u(\overline{\phi_o w_1 w_2,-1u})s = \overline{\overline{\phi_o w_1 w_2,-1u*s,u}}$ (TB6),

(TB4),(TB2)

$\longrightarrow \nu_o w_1 w_2 = w_1(\phi_o w_1 w_2) = w_1(\overline{\overline{\phi_o w_1 w_2,-1u*s,u}})<u\leqslant \nu_o w_1 w_2$

$\longrightarrow \perp$ (T29),(T31),(T28)

(TB8) follows from this by (T13).

$(TB9)^\alpha$ $\phi_o w_1 w_2(\nu_o w_1 w_2) = w_2(\nu_o w_1 w_2)(\phi w_1 w_2(\nu_o w_1 w_2)'(\overline{\phi_o w_1 w_2,\nu_o w_1 w_2}))$

<u>Proof:</u>

$z_o \equiv: \nu_o w_1 w_2$, $\quad r \equiv: \overline{\phi_o w_1 w_2, -1z_o}$

$s \equiv: \overline{\lambda x^o \phi_o w_1 w_2(-1z_o), z_o 0}$

$$\phi_o w_1 w_2 = \phi w_1 w_2 z_o rs \qquad \text{(TB6)}$$

$$= \langle\rangle(\overline{\overline{r,-1z_o}*s,z_o})(\phi w_1 w_2 z_o'(\overline{\overline{r,-1z_o}*s,z_o})a_o(z_o,r,s))z_o \qquad \text{(TB2), (TB8), (T28), (TB4)}$$

$$\phi_o w_1 w_2 z_o = \phi w_1 w_2 z_o'(\overline{\overline{r,-1z_o}*s,z_o})a_o(z_o,r,s)z_o \qquad \text{(T39), (T28)}$$

(1) $$= a_o(z_o,r,s) = w_2 z_o(\phi w_1 w_2 z_o'(\overline{\overline{r,-1z_o}*s,z_o})) \qquad \text{(TB3), (T6), (T14)}$$

(2) $$z_o \neq 0 \longrightarrow z_o = (\delta z_o)' > 0 \wedge s = \phi_o w_1 w_2(-1z_o) \qquad \text{(T8), (T37), (T40)}$$

$$\overline{\overline{r,-1z_o}*s,z_o} = \overline{\overline{\phi_o w_1 w_2,-1z_o}*s,z_o} = \overline{\phi_o w_1 w_2, z_o} \qquad \text{(TB4), (2), (T21), (TB5)}$$

With (1) this yields the statement.

VIII. Functional interpretation of classical $(AC)^0$-, (ωAC)-analysis with (ER)-qf and functional interpretation in the narrower sense of Heyting-analysis plus (ER)-qf, (MP), $(\underset{\to}{V})$, $(\neg\bigwedge\neg)^0$, $(\overset{\neg\neg}{\omega}AC)$ in T∪BR

Now the considerations from chapter VI are continued in keeping the previously made agreements.

(8.1) For every A (of the functional language) provable in Heyting-analysis plus (MP), $(\underset{\to}{V})$, $(\neg\bigwedge\neg)^0$, $(\overset{\neg\neg}{\omega}AC)$ there is a functional φ in T∪BR with: $_{T\cup BR}\vdash \varphi\text{-sol}(A')$.

Proof:

We have to complete (6.4) by $(\neg\bigwedge\neg)^0$, $(\overset{\neg\neg}{\omega}AC)$. - Because $(\overset{\neg\neg}{\omega}AC)' \equiv ((\overset{\neg\neg}{\omega}AC)\text{-}V\bigwedge)'$ and over Heyting-analysis $(\overset{\neg\neg}{\omega}AC)\text{-}V\bigwedge$ follows extensionally from $(\overset{\neg\neg}{\omega}AC)\text{-}\bigwedge$ by (1.1) (analogous to the reduction of $(\omega AC)\text{-}V\bigwedge$ to $(\omega AC)\text{-}\bigwedge$ in chapter I) and because by (6.4) all this carries over to functional interpretation in the narrower sense, so it suffices to consider $(\neg\bigwedge\neg)^0$, $(\overset{\neg\neg}{\omega}AC)\text{-}\bigwedge$.

$$((\neg\bigwedge\neg)^0)' \equiv (\bigwedge x^0 \neg\neg V y^\alpha \bigwedge z^\beta A_0(x,y,z,\underline{\underline{u}}) \longrightarrow \neg\neg\bigwedge x^0 V y^\alpha \bigwedge z^\beta A_0(x,y,z,\underline{\underline{u}}))'$$

$(A_0$ quantifierfree)

$$\equiv V x_1,x_2,x_3 \bigwedge y_1,y_2,y_3 \{A_0(x_2\underline{\underline{y}},y_1(x_2\underline{\underline{y}})(x_3\underline{\underline{y}}),x_3\underline{\underline{y}}(y_1(x_2\underline{\underline{y}})(x_3\underline{\underline{y}})),\underline{\underline{u}})$$
$$\longrightarrow A_0(y_2(x_1\underline{\underline{y}}),x_1\underline{\underline{y}}(y_2(x_1\underline{\underline{y}})),y_3(x_1\underline{\underline{y}}),\underline{\underline{u}})\}$$

where type/$x_1 \equiv \alpha 0(\beta(\alpha 0))(0(\alpha 0))(\alpha(\beta\alpha)0)$, type/$y_1 \equiv \alpha(\beta\alpha)0$

type/$x_2 \equiv 0(\beta(\alpha 0))(0(\alpha 0))(\alpha(\beta\alpha)0)$, type/$y_2 \equiv 0(\alpha 0)$

type/$x_3 \equiv \beta\alpha(\beta(\alpha 0))(0(\alpha 0))(\alpha(\beta\alpha)0)$, type/$y_3 \equiv \beta(\alpha 0)$

$$((\overset{\neg\neg}{\omega}AC)^\alpha\text{-}\bigwedge)' \equiv (\bigwedge x^0,\utilde{y}^\alpha \neg\neg V z^\alpha \bigwedge v^\beta A_0(x,\utilde{y},z,v,\underline{\underline{u}}) \longrightarrow$$
$$\neg\neg V x^{\alpha 0} \bigwedge y^0,v^\beta A_0(y,\utilde{xy},\utilde{xy'},v,\underline{\underline{u}}))' \quad (A_0 \text{ quantifierfree})$$
$$\equiv V x_1,x_2,\utilde{x_3},x_4 \bigwedge y_1,y_2,y_3 \{A_0(x_2\underline{\underline{y}},\utilde{x_3\underline{\underline{y}}},y_1(x_2\underline{\underline{y}})(\utilde{x_3\underline{\underline{y}}})(x_4\underline{\underline{y}}),$$
$$x_4\underline{\underline{y}}(y_1(x_2\underline{\underline{y}})(\utilde{x_3\underline{\underline{y}}})(x_4\underline{\underline{y}})),\underline{\underline{u}}) \longrightarrow$$
$$A_0(y_2(x_1\underline{\underline{y}}),\utilde{x_1\underline{\underline{y}}(y_2(x_1\underline{\underline{y}}))},x_1\underline{\underline{y}}(y_2(x_1\underline{\underline{y}}))\utilde{'},y_3(x_1\underline{\underline{y}}),\underline{\underline{u}})\}$$

where $type/x_1 \equiv \alpha 0(\beta(\alpha 0))(0(\alpha 0))(\alpha(\beta\alpha)\underset{\sim}{\alpha}0)$, $type/y_1 \equiv \alpha(\beta\alpha)\underset{\sim}{\alpha}0$

$type/x_2 \equiv 0(\beta(\alpha 0))(0(\alpha 0))(\alpha(\beta\alpha)\underset{\sim}{\alpha}0)$, $type/y_2 \equiv 0(\alpha 0)$

$\underset{\sim\sim\sim\sim\sim\sim\sim\sim\sim\sim\sim\sim\sim\sim\sim\sim\sim\sim}{type/x_3 \equiv \alpha(\beta(\alpha 0))(0(\alpha 0))(\alpha(\beta\alpha)\alpha 0)}$, $type/y_3 \equiv \beta(\alpha 0)$

$type/x_4 \equiv \beta\alpha(\beta(\alpha 0))(0(\alpha 0))(\alpha(\beta\alpha)\underset{\sim}{\alpha}0)$

If these two functional interpretations (in the narrower sense) are compared, one establishes immediately that $((\neg\Lambda\neg)^{o})'$ is contained in $((\neg\neg\omega AC)-\Lambda)'$ by formally crossing out the second argument of A_o together with all magnitudes referring to it and the successor ' in the third argument. The places in question are marked above by $\sim$-underlining. Therefore only the functional solution of $((\neg\neg\omega AC)^{\alpha}-\Lambda)'$ is carried out and by $\sim$-underlining it is indicated how this yields a solution of $((\neg\Lambda\neg)^{o})'$ by crossing out the magnitudes concerned.

For a identical fulfillment of $((\neg\neg\omega AC)^{\alpha}-\Lambda)'$ it suffices to produce functionals $\varphi_1, \varphi_2, \underset{\sim}{\varphi_3}, \varphi_4$ in T$\cup$BR with:

(a) $\varphi_2\underline{\underline{uv}} \overset{o}{=} v_2(\varphi_1\underline{\underline{uv}})$

(b) $\underset{\sim\sim\sim\sim\sim\sim\sim\sim\sim\sim\sim\sim}{\varphi_3\underline{\underline{uv}} \overset{\alpha}{=} \varphi_1\underline{\underline{uv}}(v_2(\varphi_1\underline{\underline{uv}}))}$

(c) $\varphi_1\underline{\underline{uv}}(v_2(\varphi_1\underline{\underline{uv}}))\underset{\sim}{'} \overset{\alpha}{=} v_1(\varphi_2\underline{\underline{uv}})\underset{\sim\sim\sim}{(\varphi_3\underline{\underline{uv}})}(\varphi_4\underline{\underline{uv}})$

(d) $\varphi_4\underline{\underline{uv}}(v_1(\varphi_2\underline{\underline{uv}})\underset{\sim\sim\sim}{(\varphi_3\underline{\underline{uv}})}(\varphi_4\underline{\underline{uv}})) \overset{\beta}{=} v_3(\varphi_1\underline{\underline{uv}})$

If $\varphi_2, \underset{\sim}{\varphi_3}$ are defined from φ_1 according to (a), (b), so it remains to exhibit only functionals φ_1, φ_4 in T$\cup$BR with:

($*_1$) $\varphi_1\underline{\underline{uv}}(v_2(\varphi_1\underline{\underline{uv}}))\underset{\sim}{'} \overset{\alpha}{=} v_1(v_2(\varphi_1\underline{\underline{uv}}))\underset{\sim\sim\sim\sim\sim\sim\sim\sim}{(\varphi_1\underline{\underline{uv}}(v_2(\varphi_1\underline{\underline{uv}})))}(\varphi_4\underline{\underline{uv}})$

($*_2$) $\varphi_4\underline{\underline{uv}}(\varphi_1\underline{\underline{uv}}(v_2(\varphi_1\underline{\underline{uv}}))\underset{\sim}{'}) \overset{\beta}{=} v_3(\varphi_1\underline{\underline{uv}})$

where $type/v_1 \equiv \alpha(\beta\alpha)\underset{\sim}{\alpha}0$ $\quad$ $type/\varphi_1\underline{\underline{uv}} \equiv \alpha 0$

$type/v_2 \equiv 0(\alpha 0)$ $\quad$ $type/\varphi_4\underline{\underline{uv}} \equiv \beta\alpha$

$type/v_3 \equiv \beta(\alpha 0)$

Such φ_1, φ_4 gives <u>bar recursion $(TB2)^{\alpha}$ of type α.</u>

(1) $\omega_1 uvw_1^{\alpha 0} \overset{o}{=}: (v_2 w_1)\underset{\sim}{'}$

(2) $\omega_2 uvw_2^{o}w_3^{\alpha 0\alpha} \overset{\alpha}{=}: v_1(\underset{\sim}{-1}w_2)\underset{\sim\sim\sim\sim\sim\sim}{(w_3\sigma^{\alpha}(-1w_2))}(\lambda x^{\alpha}v_3(w_3x))$

(3) $\varphi_1\underline{\underline{uv}} \overset{\alpha o}{=}: \phi_o(\omega_1\underline{\underline{uv}})(\omega_2\underline{\underline{uv}})$

(4) $\nu_1\underline{\underline{uv}} \overset{o}{=}: \nu_o(\omega_1\underline{\underline{uv}})(\omega_2\underline{\underline{uv}}) = \omega_1\underline{\underline{uv}}(\phi_o(\omega_1\underline{\underline{uv}})(\omega_2\underline{\underline{uv}})) = (v_2(\varphi_1\underline{\underline{uv}}))'$ (1),(3)

(5) $\varphi_4\underline{\underline{uv}} \overset{\beta\alpha}{=}: \lambda x^\alpha v_3(\phi(\omega_1\underline{\underline{uv}})(\omega_2\underline{\underline{uv}})(\nu_1\underline{\underline{uv}})'(\overline{\varphi_1\underline{\underline{uv}},\nu_1\underline{\underline{uv}}})x)$

From the properties of bar recursion $(TB2)^\alpha$ shown in chapter VII now $(*_1)$, $(*_2)$ can be proved for the so defined functionals φ_1, φ_4 as follows.

<u>Ad $(*_1)$:</u>

$\varphi_1\underline{\underline{uv}}(v_2(\varphi_1\underline{\underline{uv}}))' \overset{\alpha}{=} \phi_o(\omega_1\underline{\underline{uv}})(\omega_2\underline{\underline{uv}})(\nu_o(\omega_1\underline{\underline{uv}})(\omega_2\underline{\underline{uv}}))$ (3),(4)

$= \omega_2\underline{\underline{uv}}(\nu_1\underline{\underline{uv}})(\phi(\omega_1\underline{\underline{uv}})(\omega_2\underline{\underline{uv}})(\nu_1\underline{\underline{uv}})'(\overline{\varphi_1\underline{\underline{uv}},\nu_1\underline{\underline{uv}}}))$ (TB9), (4),(3)

(6) $= v_1(-1(\nu_1\underline{\underline{uv}}))(\phi(\omega_1\underline{\underline{uv}})(\omega_2\underline{\underline{uv}})(\nu_1\underline{\underline{uv}})'(\overline{\varphi_1\underline{\underline{uv}},\nu_1\underline{\underline{uv}}})\sigma$
$(-1(\nu_1\underline{\underline{uv}})))(\lambda x^\alpha v_3(\phi(\omega_1\underline{\underline{uv}})(\omega_2\underline{\underline{uv}})(\nu_1\underline{\underline{uv}})'(\overline{\varphi_1\underline{\underline{uv}},\nu_1\underline{\underline{uv}}})x))$ (2)

For the $(\neg\neg\omega AC)$-Λ-case alone holds:

(7) $-1(\nu_1\underline{\underline{uv}}) = -0'(v_2(\varphi_1\underline{\underline{uv}}))' = v_2(\varphi_1\underline{\underline{uv}}) < (v_2(\varphi_1\underline{\underline{uv}}))' = \nu_1\underline{\underline{uv}} = -1(\nu_1\underline{\underline{uv}})'$ (4), (T14),(T16)

(8) $\phi(\omega_1\underline{\underline{uv}})(\omega_2\underline{\underline{uv}})(\nu_1\underline{\underline{uv}})'(\overline{\varphi_1\underline{\underline{uv}},\nu_1\underline{\underline{uv}}})\sigma(-1(\nu_1\underline{\underline{uv}})) \overset{\alpha}{=} \overline{\varphi_1\underline{\underline{uv}},\nu_1\underline{\underline{uv}}}(-1(\nu_1\underline{\underline{uv}}))$

$= \varphi_1\underline{\underline{uv}}(-1(\nu_1\underline{\underline{uv}})) = \varphi_1\underline{\underline{uv}}(v_2(\varphi_1\underline{\underline{uv}}))$ (7),(TB3),(T4o)

(4) to (8) together yield $(*_1)$:

$\varphi_1\underline{\underline{uv}}(v_2(\varphi_1\underline{\underline{uv}}))' \overset{\alpha}{=} v_1(v_2(\varphi_1\underline{\underline{uv}}))(\varphi_1\underline{\underline{uv}}(v_2(\varphi_1\underline{\underline{uv}})))(\varphi_4\underline{\underline{uv}})$

<u>Ad $(*_2)$:</u>

$v_3(\varphi_1\underline{\underline{uv}}) \overset{\beta}{=} v_3(\phi_o(\omega_1\underline{\underline{uv}})(\omega_2\underline{\underline{uv}}))$ (3)

$= v_3(\phi(\omega_1\underline{\underline{uv}})(\omega_2\underline{\underline{uv}})(\nu_1\underline{\underline{uv}})'(\overline{\varphi_1\underline{\underline{uv}},\nu_1\underline{\underline{uv}}})(\varphi_1\underline{\underline{uv}}(\nu_1\underline{\underline{uv}})))$ (TB7),(3)

$= \varphi_4\underline{\underline{uv}}(\varphi_1\underline{\underline{uv}}(\nu_1\underline{\underline{uv}})) = \varphi_4\underline{\underline{uv}}(\varphi_1\underline{\underline{uv}}(v_2(\varphi_1\underline{\underline{uv}}))')$ (5),(4)

Thus the functional solution of $((\neg\Lambda\neg)^o\text{-}\vee y^\alpha\Lambda z^\beta)'$, $((\neg\neg\omega AC)^\alpha\text{-}\Lambda v^\beta)'$ can be given for <u>arbitrary β</u> by bar recursion $(TB2)^\alpha$ of type α. - The whole proceeding is independent of A_o and the free variables $\underline{\underline{u}}$ in A_o.

(8.2) Classical $(AC)^o$- and (ωAC)-analysis is functional interpretable in T$\cup$BR. Accordingly now (4.5) applies. - For $(AC)^{o,\alpha}-\bigwedge z^\beta$, $(\omega AC)^\alpha-\bigwedge v^\beta$ this functional interpretation uses for arbitrary β only bar recursion of type α. For $(C)^o-\underset{\bigwedge}{\bigvee} x^\alpha \underset{\bigvee}{\bigwedge} y^\beta$, $\alpha\equiv 0\alpha_1\ldots\alpha_m$ these considerations give a functional interpretation by bar recursion of type $\beta\alpha\alpha_1\ldots\alpha_m$, what however can often be further reduced extensionally. So especially for Π^1_1-, Σ^1_1-analysis ($\alpha\equiv 00$, $\beta\equiv 0$) bar recursion of type 0(00) is needed. - To this all further cases of $(AC)^o$, (ωAC), $(C)^o$ can be reduced by (1.1), (1.2); however by **functional** contractions in general types are raised.

<u>Proof:</u> (8.1),(3.11),(3.12). - The type statements for $(AC)^{o,\alpha}-\bigwedge$, $(\omega AC)^\alpha-\bigwedge$ result directly from these proofs. -

$(C)^o-\underset{\bigwedge}{\bigvee} x^\alpha \underset{\bigvee}{\bigwedge} y^\beta$ follows classically with $(AC)^o$ from

$$\bigwedge u^o \bigvee v^o (v=0 \longleftrightarrow \underset{\bigwedge}{\bigvee} x^\alpha \underset{\bigvee}{\bigwedge} y^\beta A_o(x,y,u)) \longleftrightarrow \bigwedge u^o \bigvee v^o \{\bigvee x_1^\alpha \bigwedge y_1^\beta \ldots \wedge \bigwedge x_2^\alpha \bigvee y_2^\beta \ldots\}$$

$$\longleftrightarrow \bigwedge u^o \bigvee v^o, x_1^\alpha, y_2^{\beta\alpha} \bigwedge y_1^\beta, x_2^\alpha \ldots \longleftrightarrow \bigwedge u^o \bigvee z^{\beta\alpha\alpha_1\ldots\alpha_m} \bigwedge x^{\gamma o} \ldots$$

((Tnd),(AC)-qf, extensional quantifier contraction chapter I)

The now needed $(AC)^{o,\beta\alpha\alpha_1\ldots\alpha_m}-\bigwedge$ then requires for functional interpretation bar recursion of type $\beta\alpha\alpha_1\ldots\alpha_m$ or of a type reduced from it. Especially for $\alpha\equiv 00$, $\beta\equiv 0$ this gives at first 0(00)0, what after chapter I can be extensionally reduced to 0(00).

Concluding we shortly mention a more intensional treatment. Extensionality was employed in the results of this chapter firstly for functional contractions and secondly in the derivations of chapter VII. The sequence-extensionality used in chapter VII is so not eliminable. The other extensionality applications at the functional contractions can be avoided by using tupels. However this leads to the following changes.

Whereas we continue working with one bar ω_1 there may now be several (homogeneous) arguments. Such a manyfold bar recursion has to produce the further (homogeneous) magnitudes corresponding to the arguments in a simultaneous and symmetrical way. One convinces oneself that this is performed by the (TB2)-analogon of the following <u>simultaneous bar</u>

<u>recursion</u> (corresponding to (TB1))

$$\phi_i w_1 \underline{\underline{w}}_2 \underline{\underline{w}}_3 u^o \underline{\underline{v}} = \begin{cases} w_{2i} & \text{if } w_1(\underline{\underline{<v_j 0, \ldots, v_j(u-1)>}}) < u \\ w_{3i}(\underline{\underline{\lambda x_1, \ldots, x_n \phi_k w_1 \underline{\underline{w}}_2 \underline{\underline{w}}_3 u' <v_j 0, \ldots, v_j(u-1), x_j>}}) & \text{otherwise} \end{cases}$$

$$\left.\begin{aligned} \text{where } \underline{\underline{w}}_2 &\equiv w_{21}, \ldots, w_{2m} \\ \underline{\underline{w}}_3 &\equiv w_{31}, \ldots, w_{3m} \end{aligned}\right\} \quad i,k = 1, \ldots, m$$

$$\underline{\underline{v}} \equiv v_1, \ldots, v_n \qquad j = 1, \ldots, n$$

So analysis can be handled intensionally up to sequence-extensionality.- However the extensional treatment simplifies the situation considerably. Moreover the extensional character of the means used is explicitly stated.

IX. Further consequences from the functional interpretation of classical analysis

1. Arithmetical comprehension and Δ-comprehension over type 0

As well known A is a Δ-predicate iff it is equivalent to prenex formulae with dual quantifier structure. Here it suffices to consider the following principal case of Δ-comprehension over numbers:

$$(C)^{0}\text{-}\Delta^{\alpha,\beta}:\ V \rightarrow \bigvee z^{00} \bigwedge x^{0}\{zx=0 \longleftrightarrow \bigvee x_1^{\alpha} \bigwedge y_1^{\beta} A_0(x_1,y_1,x)\}$$
$$\text{where } V \equiv:\ \bigwedge x^{0}(\bigvee x_1^{\alpha} \bigwedge y_1^{\beta} A_0(x_1,y_1,x) \longleftrightarrow \bigwedge x_2^{\alpha} \bigvee y_2^{\beta} B_0(x_2,y_2,x))$$
$$(A_0, B_0 \text{ quantifierfree})$$

(9.1) The given functional interpretation covers $(C)^{0}\text{-}\Delta^{\alpha,\beta}$ by bar recursion of type α independently of β. Because arithmetical predicates (only quantifiers of type 0) are Δ_1^1 so arithmetical and hyper-arithmetical comprehension of sets of numbers is functional interpretated by bar recursion of type 00. - To $(C)^{0}\text{-}\Delta^{\alpha,\beta}$ all other Δ-comprehensions over type 0 reduce by (1.1), (1.2). For instance $(C)^{0}\text{-}\Delta_2^1$ comes in this way under bar recursion of the same type as $(C)^{0}\text{-}\Pi_1^1,\Sigma_1^1$, namely 0(00).

Proof:
$(C)^{0}\text{-}\Delta^{\alpha,\beta}$ follows in classical analysis from

$$V \longrightarrow \bigwedge x^{0} \bigvee z^{0}\{z=0 \longleftrightarrow \bigvee x_1^{\alpha} \bigwedge y_1^{\beta} A_0(x_1,y_1,x)\}$$
$$\longleftrightarrow \bigwedge x^{0} \bigvee z^{0}, x_1^{\alpha}, x_2^{\alpha} \bigwedge y_1^{\beta}, y_2^{\beta} \ \ldots \qquad \text{(V-def.)}$$
$$\longleftrightarrow \bigwedge x^{0} \bigvee x_3^{\alpha} \bigwedge y_3^{\beta} \ \ldots \qquad \text{(quantifier contraction)}$$

by (Tnd), $(AC)^{0,\alpha}\text{-}\bigwedge y_3^{\beta}$. For this the above functional interpretation uses after (8.2) bar recursion of type α.

For arithmetical predicates one easily proves by prenexation and induction on the numbers of quantifiers with $(AC)^{0,00}\text{-}\bigwedge z^{0}$ and functional contraction the equivalence to $\bigvee x^{00} \bigwedge y^{0}$-formulae. This applied to the negation gives directly that arithmetical predicates are Δ_1^1. After (8.2) and the preceding therefore arithmetical and hyper-arithmetical comprehension $(C)^{0}\text{-}\Delta_1^1 \equiv (C)^{0}\text{-}\Delta^{00,0}$ fall under the functional interpretation with bar recursion of type 00.

For $(C)^{0}\text{-}\Delta_2^1$ from this the reduction according to (1.1),(1.2) gives on account of $\bigvee x^{00} \bigwedge y^{00} \bigvee z^{0} \ldots \longleftrightarrow \bigvee x^{00}, z_1^{0(00)} \bigwedge y^{00} \ldots \longleftrightarrow \bigvee z^{0(00)} \bigwedge y^{00} \ldots$

a functional interpretation with bar recursion of type 0(00).

2. Double negation of the tertium-non-datur over numbers

Because of the intuitionistic deductive equivalence of $(^{\neg}\bigwedge^{\neg})^{o}$ and $\neg\neg\bigwedge x^{o}(\mathrm{Tnd})$ (see chapter III) the functional interpretation of analysis gives also directly a functional interpretation in the narrower sense of the consistency statement for the tertium-non-datur on numbers resp. for the equivalent (3.5) type 0-comprehension. We will explain this in more detail for $(\neg\neg\bigwedge x^{o}(\mathrm{Tnd}))'$.

The functional interpretation of the intuitionistic $\bigwedge x^{o}\neg\neg(\mathrm{Tnd})$

$$(\bigwedge x^{o}\neg\neg(\mathrm{Tnd}))' \equiv (\bigwedge x^{o}\neg\neg\{\bigvee y_1^{\alpha}\bigwedge z_1^{\beta}A_o(y_1,z_1,x,\underline{u}) \vee \neg\bigvee y_2^{\alpha}\bigwedge z_2^{\beta}A_o(y_2,z_2,x,\underline{u})\})'$$

$(A_o$ quantifierfree)

$$\equiv \bigvee x_1,x_2,x_3 \bigwedge y_1^{o}, y_2^{\beta(\beta\alpha)\alpha o}, y_3^{\alpha(\beta\alpha)\alpha o}\{(x_1\underline{y}=0 \wedge$$

$$A_o(x_2\underline{y}, y_2(x_1\underline{y})(x_2\underline{y})(x_3\underline{y}), y_1, \underline{u})) \vee (x_1\underline{y}\neq 0 \wedge$$

$$\neg A_o(y_3(x_1\underline{y})(x_2\underline{y})(x_3\underline{y}), x_3\underline{y}(y_3(x_1\underline{y})(x_2\underline{y})(x_3\underline{y})), y_1, \underline{u}))\}$$

where $\mathrm{type}/x_1 \equiv 0(\alpha(\beta\alpha)\alpha 0)(\beta(\beta\alpha)\alpha 0)0$

$\mathrm{type}/x_2 \equiv \alpha(\alpha(\beta\alpha)\alpha 0)(\beta(\beta\alpha)\alpha 0)0$

$\mathrm{type}/x_3 \equiv \beta\alpha(\alpha(\beta\alpha)\alpha 0)(\beta(\beta\alpha)\alpha 0)0$

has the following primitive recursive functional solution depending (by (4.6) necessarily) on the prime formulae:

$$B(\underline{u},\underline{v}) \equiv: \neg A_o(v_3 1\sigma^{\alpha}(\lambda x^{\alpha}v_2 0x\sigma^{\beta\alpha}), v_2 0(v_3 1\sigma^{\alpha}(\lambda x^{\alpha}v_2 0x\sigma^{\beta\alpha}))\sigma^{\beta\alpha}, v_1, \underline{u})$$

$$\varphi_1\underline{u}\underline{v}\ ^{\underline{o}}: \begin{cases} 1 & \text{if } B(\underline{u},\underline{v}) \\ 0 & \text{otherwise} \end{cases} \qquad \varphi_2\underline{u}\underline{v}\ ^{\underline{\alpha}}: \begin{cases} \sigma & \text{if } B(\underline{u},\underline{v}) \\ v_3 1\sigma^{\alpha}(\lambda x^{\alpha}v_2 0x\sigma^{\beta\alpha}) & \text{otherwise} \end{cases}$$

$$\varphi_3\underline{u}\underline{v}\ ^{\underline{\beta\alpha}}: \begin{cases} \lambda x^{\alpha}v_2 0x\sigma^{\beta\alpha} & \text{if } B(\underline{u},\underline{v}) \\ \sigma & \text{otherwise} \end{cases}$$

This yields with the functional solution of $(\bigwedge x^{o}\neg\neg(\mathrm{Tnd}) \longrightarrow \neg\neg\bigwedge x^{o}(\mathrm{Tnd}))'$ from chapter VIII, which is independent of the prime formulae, on account of (6.1), (MP) a functional solution of $(\neg\neg\bigwedge x^{o}(\mathrm{Tnd}))'$ depending (by (4.6) necessarily) on the prime formulae. - Because of its independence of the prime formulae the functional interpretation (in the narrower sense) of $(^{\neg}\bigwedge^{\neg})^{o}$ is simpler than that of $\neg\neg\bigwedge x^{o}(\mathrm{Tnd})$, $\neg\neg(C)^{o}$. Analogical the functional interpretation of axioms of choice is simpler than that of comprehension axioms $(\neg\neg((C)^{o}\text{-}V)^{*} \leftrightarrow \neg\neg((C)^{o}\text{-}V)$, (MP), (3.15), (4.6)).

3. Church-thesis

In [19], 330 Howard and Kreisel show intuitionistically the incompatibility of Church-thesis and $\bigwedge x^{o}(\mathrm{Tnd})$. For this we give at first a more direct proof and then by the help of the functional interpretation of analysis we draw from it some remarkable consequences.

Let $T_1(u^{o},v^{o},w^{o})$ be the primitive recursive predicate and Uv^{o} the primitive recursive function from Kleene [17], 281,288. Using the notation of [17] we have:

(1) $T_1(u,v,w) \rightarrow S_1(u,v,w) \rightarrow lh((w)_{0,1})\neq 0 \rightarrow (w)_{0,1}\neq 0 \rightarrow w\neq 0$ (p.278)

(2) $T_1(u,v,w_1) \wedge T_1(u,v,w_2) \rightarrow w_1=w_2$

$Uv = \mu x_{x<(v)_{0,2}} Nu((v)_{0,2},x)\leq(v)_{0,2}$ (p.278)

$U0 = 0, \quad v\neq 0 \rightarrow Uv<v$

(3) $Uv=v \leftrightarrow v=0$

As outlined these relations (1) - (3) can be proved in Heyting-arithmetic.

By Church-thesis we understand the following general recursive axiom of choice:

(ChT) $\bigwedge x^{o}\bigvee y^{o}A(x,y) \rightarrow \bigvee e^{o}\bigwedge x^{o}\bigvee z^{o}\{T_1(e,x,z) \wedge A(x,Uz)\}$

For the following applications of the tertium-non-datur resp. Church's thesis

$A \equiv: \bigwedge x^{o}\{\bigvee y^{o}T_1(x,x,y) \vee \neg\bigvee y_1 T_1(x,x,y_1)\}$

$B \equiv: \bigwedge x^{o}\bigvee y^{o}\{T_1(x,x,y) \vee \neg\bigvee y_1 T_1(x,x,y_1)\} \rightarrow \bigvee e^{o}\bigwedge x^{o}\bigvee z^{o}\{T_1(e,x,z) \wedge (T_1(x,x,Uz) \vee \neg\bigvee y_1 T_1(x,x,y_1))\}$

in <u>Heyting-arithmetic</u> now holds:

$A \wedge B \rightarrow \bigvee e\bigwedge x\bigvee z\{T_1(e,x,z) \wedge (T_1(x,x,Uz) \vee \bigwedge y_1\neg T_1(x,x,y_1))\}$

$\rightarrow \bigvee e,z\{T_1(e,e,z) \wedge T_1(e,e,Uz)\}$ (x:e)

$\rightarrow \bigvee z\{z\neq 0 \wedge z=Uz\}$ (1),(2)

$\rightarrow \perp$ (3)

Hence with $\neg(C\rightarrow D) \leftrightarrow \neg\neg C \wedge \neg D$: (4) $\neg\neg A \leftrightarrow \neg B$

Church-thesis and tertium-non-datur are intuitionistically incompatible.

(4),(6.2) yield: $_T\vdash \varphi\text{-sol}(\neg\neg A \leftrightarrow \neg B)'$. After chapters VIII, IX.2 $(\neg\neg A)'$ has a functional solution in T∪BR by bar recursion of type 0: $T\cup(BR)^{o}\cup(\omega) \vdash \varphi\text{-sol}(\neg\neg A)'$. With (6.1), (MP) now follows:

(9.2) $T\cup(BR)^{o}\cup(\omega) \vdash \varphi\text{-sol}(\neg B)'$. The negation of an application of Church-thesis is functional interpretable (in the narrower sense) in $T\cup(BR)^{o}\cup(\omega)$. Thus (ChT) is in no extension of $T\cup(BR)^{o}\cup(\omega)$ and therefore also not intuitionistically consistent functional interpretable (in the narrower sense).

The contrariness of (ChT) and functional interpretation in the narrower sense can still be more localized.

$$\neg B \leftrightarrow \neg\neg \wedge x \vee y\{T_1(x,x,y) \vee \wedge y_1 \neg T_1(x,x,y_1)\} \wedge \neg \vee e \wedge x \vee z\{T_1(e,x,z) \wedge (T_1(x,x,Uz) \vee \wedge y_1 \neg T_1(x,x,y_1))\}$$

$$(5) \leftrightarrow \neg\neg \wedge x \vee y \wedge y_1 A_o(x,y,y_1) \wedge \neg \vee e \wedge x \vee z \wedge y_1 B_o(e,x,z,y_1)$$

$(A_o, B_o$ quantifierfree)

By (9.2), (6.1) we therefore have

$$(6) \quad T\cup(BR)^{o}\cup(\omega) \vdash \varphi\text{-sol}(\neg\neg \wedge x \vee y \wedge y_1 A_o(x,y,y_1))' \wedge \varphi\text{-sol}(\neg \vee e \wedge x \vee z \wedge y_1 B_o(e,x,z,y_1))'$$

Now we pass over to <u>Heyting-analysis plus $(BR)^{o}$, (ω)</u>. After (6) we can prove herein:

$$(7) \quad (\neg\neg \wedge x \vee y \wedge y_1 A_o(x,y,y_1))' \wedge (\neg \vee e \wedge x \vee z \wedge y_1 B_o(e,x,z,y_1))'$$

$$(\neg \vee e \wedge x \vee z \wedge y_1 B_o(e,x,z,y_1))' \equiv \vee x^{o(oo)o}, y_1^{o(oo)o} \wedge e^{o}, z^{oo}\ \neg B_o(e,xez,z(xez),y_1 ez)$$

$$\rightarrow \wedge e^{o}, z^{oo} \vee x^{o}, y_1^{o} \neg B_o(e,x,zx,y_1)$$

$$\rightarrow \neg \vee e, z^{oo} \wedge x, y_1 B_o(e,x,zx,y_1)$$

$$(8) \qquad \rightarrow \neg \vee e \wedge x \vee z^{o} \wedge y_1 B_o(e,x,z,y_1)$$

$$(\neg\neg \wedge x \vee y \wedge y_1 A_o(x,y,y_1))' \equiv \vee y^{oo(o(oo))(o(oo))} \wedge x^{o(oo)}, y_1^{o(oo)}\ A_o(x(yxy_1), yxy_1(x(yxy_1)), y_1(yxy_1))$$

$$\rightarrow \wedge x^{o(oo)}, y_1^{o(oo)} \vee y^{oo} \neg\neg A_o(xy, y(xy), y_1 y)$$

$(\neg\neg\bigwedge x \bigvee y \bigwedge y_1 A_o(x,y,y_1))' \rightarrow \neg\bigvee x^{o(oo)}, y_1^{o(oo)} \bigwedge y^{oo} \neg A_o(xy, y(xy), y_1 y)$

(9) $\rightarrow \neg\bigwedge y^{oo} \bigvee x^o, y_1^o \neg A_o(x, yx, y_1)$

Consider now the following application of $(\neg\bigwedge\neg)^{oo}$-$\bigvee x^o$:

$C \equiv: (\bigwedge y^{oo} \neg\neg\bigvee x^o, y_1^o \neg A_o(x,yx,y_1) \rightarrow \neg\neg\bigwedge y \bigvee x, y_1 \neg A_o(x,yx,y_1))$

$C \wedge \neg\bigwedge y^{oo} \bigvee x^o, y_1^o \neg A_o(x,yx,y_1) \rightarrow \neg\neg\bigvee y^{oo} \bigwedge x^o, y_1^o A_o(x,yx,y_1)$

(10) $\rightarrow \neg\neg\bigwedge x^o \bigvee y^o \bigwedge y_1^o A_o(x,y,y_1)$

(5) to (10) yield: $C \rightarrow \neg B$

The Markov-principle for numbers and 00-parameter obviously implies C. Consequently the following is proved:

(9.3) Church's thesis is intuitionistically (in Heyting-analysis plus $(BR)^o$, (ω)) with the consistency statement of the Markov-principle for numbers and 00-parameter and even with $(\neg\bigwedge\neg)^{oo}$-$\bigvee x^o$ incompatible.

X. Consistency proof by computation. Computation of $T\cup BR^{0\ldots 0}{}_{\cup\mu}$

The basic idea of functional interpretation is a reduction to constructions. The previously presented reduction of classical number theory and analysis to T resp. T∪BR gives after (4.5) to the original statements a certain pure functional meaning which among other things carries over consistency. In order to consider this as constructive content we have to prove T, T∪BR as calculi of constructions. So we have to show that <u>within</u> these calculi each functional can be computed consistently, that is there exists at least one computation and all other computations give the same result. We prove this in two steps. At first a standard computation is layed down and with the help of step notions the standard computability of all functionals in the calculi T, T∪BR is shown. Afterwards it is proved that relative to this computation only "valid" formulae are derivable in the calculi.

In order to avoid frequent meta-deliminations in these meta-considerations we use the familiar mathematical language.

The computation of a closed term of type 0 can be arranged in a natural manner as a spread at the nodes of which are the subterms needed in the computation. For the branchings the following form can be fixed:

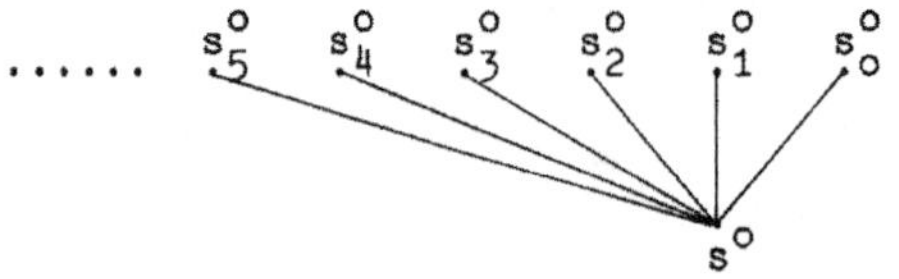

s_i are the immediate reduction terms of s; mostly the computation of s_j, j>i precedes that of s_i. Thus s_0 is the main result of the whole reduction step reflected by this ramification. - Such a <u>computation tree</u> formed according to the computation instructions represents a calculation if it contains at each node only finitely many ramifications and if each string in it breaks off after finitely many steps. By the fan theorem such a tree comprises exactly in this case finitely many computation steps. The <u>value of each subterm</u> in the tree stands always at the end of the string starting with the term and then proceeding entirely to the right.

Now we fix the here needed instructions for a natural standard computation. Because in the following we operate only with this standard computation we suppress constantly the supplement "standard".

We investigate several successive expanding functional domains $\mathcal{f}_i$. The terms of each functional domain are built up from the functional constants by application; consequently they are closed. To the operations of T∪BR here we still add choice sequences $<r_0^\alpha, r_1^\alpha, \ldots>^{\alpha 0}$ and a certain μ-operation. Whereas at the beginning to these functional domains (e.g. T) a complete system of notations can effectively be given, later on after the addition of all choice sequences of a type (e.g. over numbers) this is of course impossible. But to all functional domains here considered intuitionistically an existence is adjudged.

Reduction instructions for the (standard-)computability

Consider a functional domain $\mathcal{f}$ and terms $s^0, \varphi, t_1, \ldots, t_p, r_0, r_1, \ldots \in \mathcal{f}$ of suitable types: $s^0 \equiv \varphi t_1 \ldots t_p$, φ functional constant.

0. Only numerals have no reduction terms.

The reduction instructions for other s^0 depend on the first functional constant in s.

1. $\varphi \equiv '$: p=1
s has the two reduction terms $s_1^0 \equiv: t_1^0$, $s_0^0 \equiv:$ (value/s_1)'.

2. φ according to (T1): $\varphi u_1 \ldots u_n \overset{\alpha}{=} u_1$, $1 \leqslant n \leqslant p$
s has the sole reduction term $s_0^0 \equiv: t_1 t_{n+1} \ldots t_p$.

3. φ according to (T2): $\varphi u_1 \ldots u_n \overset{\alpha}{=} \psi$ (ψ preceding resp. 0,') $0 \leqslant n \leqslant p$

3.1 $\psi \equiv 0$: $\alpha \equiv 0$, p=n
s has the only reduction term $s_0^0 \equiv: 0$.

3.2 $\psi \equiv '$: $\alpha \equiv 00$, p=n+1
s has the two reduction terms $s_1^0 \equiv: t_p^0$, $s_0^0 \equiv:$ (value/s_1)'.

3.3 ψ preceding, $\not\equiv$ 0, ':
s has the sole reduction term $s_0^0 \equiv: \psi t_{n+1} \ldots t_p$.

4. φ according to (T3): $\varphi u_1 \ldots u_n \overset{\alpha}{=} \psi u_1 \ldots u_n (\chi u_1 \ldots u_n)$ (ψ, χ preceding) $0 \leqslant n \leqslant p$

s has the only reduction term $s_o^o \equiv: \psi t_1 \ldots t_n(\chi t_1 \ldots t_n)t_{n+1} \ldots t_p$.

5. φ according to (T4): $\begin{cases} \varphi 0 \overset{\alpha}{=} \psi \\ \varphi u' = \chi(\varphi u)u^o \end{cases}$ $(\psi,\chi$ preceding)

5.1 t_1^o is a numeral.

5.1.1: $t_1 \equiv 0$

s has the sole reduction term $s_o^o \equiv: \psi t_2 \ldots t_p$.

5.1.2: $t_1 \equiv z'$ (numeral z)

s has the only reduction term $s_o^o \equiv: \chi(\varphi z)zt_2 \ldots t_p$.

5.2 t_1^o is no numeral.

s has the two reduction terms $s_1^o \equiv: t_1^o$, $s_o^o \equiv: \varphi(\text{value}/s_1)t_2 \ldots t_p$.

6. φ according to $(BR)^\alpha$: $p \geq 6$

$w_1(\overline{\overline{v,-1u*w,u}})<u \longrightarrow \phi w_1^{o(\alpha o)} w_2 w_3 u^o v^{\alpha o} w^\alpha = w_2$

$w_1(\overline{\overline{v,-1u*w,u}}) \geq u \longrightarrow \phi w_1 w_2 w_3 uvw = w_3(\phi w_1 w_2 w_3 u'(\overline{\overline{v,-1u*w,u}}))$

6.1 t_4^o is a numeral z.

s has the two reduction terms $s_1^o \equiv: t_1(\overline{\overline{t_5,-1z*t_6,z}})$ and

if $\text{value}/s_1 < z$, so $s_o^o \equiv: t_2 t_7 \ldots t_p$

if $\text{value}/s_1 \geq z$, so $s_o^o \equiv: t_3(\phi t_1 t_2 t_3 z'(\overline{\overline{t_5,-1z*t_6,z}}))t_7 \ldots t_p$.

6.2 t_4^o is no numeral.

s has the two reduction terms $s_1^o \equiv: t_4^o$,

$s_o^o \equiv: \phi t_1 t_2 t_3(\text{value}/s_1)t_5 t_6 t_7 \ldots t_p$.

7. φ is choice sequence $\langle r_o^\alpha, r_1^\alpha, \ldots \rangle^{\alpha o}$ over $\mathcal{F}$:

7.1 t_1^o is a numeral z.

s has the only reduction term $s_o^o \equiv: r_z t_2 \ldots t_p$.

7.2 t_1^o is no numeral.

s has the two reduction terms $s_1^o \equiv: t_1$, $s_o^o \equiv: \varphi(\text{value}/s_1)t_2 \ldots t_p$.

8. φ is μ-operator:

Our considerations here refer to the following μ-formation:

$\mu w^{o(\alpha o)} v^{\alpha o} \overset{o}{=}: \mu x^{o}\{w(\langle v0,\dots,v(x-1)\rangle)<x\}$, where $\langle v0,\dots,v(x-1)\rangle$ (finite choice sequence) denotes for $x\neq 0$ the choice sequence $\langle v0,\dots,v(-1x),\sigma,..\rangle$ and for $x=0$ the choice sequence $\langle\sigma,\dots\rangle$. So this functional μ is characterized in a functional calculus by the axioms

$w(\langle v0,\dots,v(u-1)\rangle)<u^{o} \rightarrow \mu wv\leq u, \quad w(\langle v0,\dots,v(\mu wv-1)\rangle)<\mu wv.$

Accordingly for $s^{o} \equiv \mu t_1 t_2$ (p=2) it is fixed:

For $i\geq 1$: $s_i^{o} \equiv: t_1(\langle t_2 0,\dots,t_2(i-1)\rangle)$ if $\neg\bigvee 1\leq j<i$ value/$s_j<j$

s_o^{o} be the number of these $s_1^{o},s_2^{o},\dots$ provided it is finite. Otherwise such a tree part belongs to no bounded computation tree and a s_o-specification is superfluous.

1. - 8. are within a functional calculus unambiguous instructions. Our aim is to show that for the functional domains $\mathcal{F}$ considered these reductions yield to every type 0-term exactly one finite computation tree.

Now by type induction the fundamental notions "extensionally computable" (e-comp) and "extensional computational equal" ($\overset{\alpha}{=}_{ec}$) in each case relative to $\mathcal{F}$ are defined.

Definition: Consider terms of $\mathcal{F}$.

1. t^{o} e-comp (in $\mathcal{F}$) $\Longleftrightarrow$: computation tree of t is finite

   $t \overset{o}{=}_{ec} s$ (over $\mathcal{F}$) $\Longleftrightarrow$: t,s e-comp (in $\mathcal{F}$) to the same value

2. t^{α} e-comp (in $\mathcal{F}$) $\Longleftrightarrow$: $\bigwedge \underline{\underline{r}},\underline{\underline{s}}$ e-comp (in $\mathcal{F}$) $\{\bigwedge_i r_i \overset{=}{=}_{ec} s_i$ (over $\mathcal{F}$) $\Longrightarrow$ $t\underline{\underline{r}} \overset{o}{=}_{ec} t\underline{\underline{s}}$ (over $\mathcal{F}$)$\}$

   $t \overset{\alpha}{=}_{ec} s$ (over $\mathcal{F}$) $\Longleftrightarrow$: t,s e-comp (in $\mathcal{F}$) $\wedge \bigwedge \underline{\underline{r}}$ e-comp (in $\mathcal{F}$) $t\underline{\underline{r}} \overset{o}{=}_{ec} s\underline{\underline{r}}$ (over $\mathcal{F}$)

At first some direct conclusions from these concepts.

(10.1) $t \overset{=}{=}_{ec} t \Longleftrightarrow t$ e-comp, $s \overset{=}{=}_{ec} t \Longrightarrow t \overset{=}{=}_{ec} s$, $s \overset{=}{=}_{ec} t \wedge r \overset{=}{=}_{ec} s \Longrightarrow r \overset{=}{=}_{ec} t$,

$s \overset{=}{=}_{ec} t \Longleftrightarrow \bigwedge \underline{\underline{u}},\underline{\underline{v}}\{\bigwedge_i u_i \overset{=}{=}_{ec} v_i \Longrightarrow s\underline{\underline{u}} \overset{o}{=}_{ec} t\underline{\underline{v}}\}$

(10.2) Numerals, ', σ are e-comp in $\mathcal{F}$.

(10.3) A choice sequence over in $\mathcal{F}$ e-comp objects (e.g. numerals) is e-comp in $\mathcal{F}$.

Proof: Under the assumption $t_o^\alpha, t_1^\alpha, \ldots, \underline{r}, \underline{s}$ e-comp we have to show:

$$\bigwedge_i r_i \equiv_{ec} s_i \Longrightarrow \langle t_o,\ldots\rangle^{\alpha o}\underline{r} \overset{\circ}{=}_{ec} \langle t_o,\ldots\rangle\underline{s}$$

As $r_o \overset{\circ}{=}_{ec} s_o \overset{\circ}{=}_{ec} z$ (numeral), so $\langle t_o,\ldots\rangle r_o^o r_1\ldots r_n \overset{\circ}{=}_{ec} \langle t_o,\ldots\rangle z r_1\ldots r_n$ $\equiv_{ec} t_z r_1 \ldots r_n \equiv_{ec} t_z s_1\ldots s_n \equiv_{ec} \langle t_o,\ldots\rangle z s_1\ldots s_n \equiv_{ec} \langle t_o,\ldots\rangle s_o s_1\ldots s_n$.

(10.4) Choice sequences over choice sequences over ... over choice sequences over numerals (iterated choice sequences over numerals) are e-comp in $\mathcal{F}$.

Proof: (10.2), (10.3)

(10.5) $\{\varphi\}$ e-comp (in $\mathcal{F}$) $\wedge \bigwedge_i r_i \equiv_{ec} s_i$ (over $\mathcal{F}$) $\Longrightarrow t(\underline{r}) \equiv_{ec} t(\underline{s})$ (over $\mathcal{F}$)

where $\{\varphi\}$ with the exception of choice sequences are the constants in $t \varepsilon \mathcal{F}$.

Proof: Induction over the term tree $\mathfrak{T}$ of t beginning with $\underline{r}$

I. $\mathfrak{T}$ unramified:

Case 1: $t(\underline{r}) \equiv \varphi$ Conclusion from the premise with (10.1).

Case 2: $t(\underline{r}) \equiv r_i$ trivial

II. $\mathfrak{T}$ ramified:

Case 1: $t(\underline{r}) \equiv t_1(\underline{r})(t_2(\underline{r}))$

Ind.hyp.: (1) $\{\varphi\}$ e-comp $\wedge \bigwedge_i r_i \equiv_{ec} s_i \Longrightarrow t_j(\underline{r}) \equiv_{ec} t_j(\underline{s})$ $(j=1,2)$

We have to show:

$(*_1)$ $\{\varphi\}$ e-comp $\wedge \bigwedge_i r_i \equiv_{ec} s_i \Longrightarrow t_1(\underline{r})(t_2(\underline{r})), t_1(\underline{s})(t_2(\underline{s}))$ e-comp

$(*_2)$ $\{\varphi\}$ e-comp $\wedge \bigwedge_i r_i \equiv_{ec} s_i \wedge \underline{w}$ e-comp $\Longrightarrow t_1(\underline{r})(t_2(\underline{r}))\underline{w} \overset{\circ}{=}_{ec} t_1(\underline{s})(t_2(\underline{s}))\underline{w}$

Ad $(*_1)$:

$\{\varphi\}$ e-comp $\wedge \bigwedge_i r_i \equiv_{ec} s_i \Longrightarrow t_1(\underline{r}), t_2(\underline{r})$ e-comp $\wedge\, t_2(\underline{r}) \equiv_{ec} t_2(\underline{r})$ (1), (10.1)

$$\Longrightarrow \bigwedge_k w_{1k} \equiv_{ec} w_{2k} \Longrightarrow t_1(\underline{r})(t_2(\underline{r}))\underline{w}_1 \overset{\circ}{=}_{ec} t_1(\underline{r})(t_2(\underline{r}))\underline{w}_2$$

$$\Longrightarrow t_1(\underline{r})(t_2(\underline{r})) \text{ e-comp}$$

Analogous for the second statement in $(*_1)$.

Ad $(*_2)$:

$$\{\varphi\} \text{ e-comp} \wedge \bigwedge_i r_i \equiv_{ec} s_i \wedge \underline{w} \text{ e-comp} \Longrightarrow t_1(\underline{r})(t_2(\underline{s}))\underline{w} \overset{\circ}{=}_{ec} t_1(\underline{s})(t_2(\underline{s}))\underline{w} \wedge$$

$$t_1(\underline{r})(t_2(\underline{r}))\underline{w} \overset{\circ}{=}_{ec} t_1(\underline{r})(t_2(\underline{s}))\underline{w}$$

(1),(10.1)

$$\Longrightarrow t_1(\underline{r})(t_2(\underline{r}))\underline{w} \overset{\circ}{=}_{ec} t_1(\underline{s})(t_2(\underline{s}))\underline{w}$$

Case 2: $t(\underline{r}) \equiv \langle t_o(\underline{r}), t_1(\underline{r}), \ldots \rangle$

Ind.hyp.: (2) $\{\varphi\} \text{ e-comp} \wedge \bigwedge_i r_i \equiv_{ec} s_i \Longrightarrow t_j(\underline{r}) \equiv_{ec} t_j(\underline{s}) \quad (j=0,1,\ldots)$

We have to show:

$(*_3)$ $\{\varphi\} \text{ e-comp} \wedge \bigwedge_i r_i \equiv_{ec} s_i \Longrightarrow \langle t_o(\underline{r}), \ldots \rangle, \langle t_o(\underline{s}), \ldots \rangle \text{ e-comp}$

$(*_4)$ $\{\varphi\} \text{ e-comp} \wedge \bigwedge_i r_i \equiv_{ec} s_i \wedge \underline{w} \text{ e-comp} \Longrightarrow \langle t_o(\underline{r}), \ldots \rangle \underline{w} \overset{\circ}{=}_{ec} \langle t_o(\underline{s}), \ldots \rangle \underline{w}$

Ad $(*_3)$:

$$\{\varphi\} \text{ e-comp} \wedge \bigwedge_i r_i \equiv_{ec} s_i \wedge \bigwedge_k w_{1k} \equiv_{ec} w_{2k} \Longrightarrow w_{1o} \overset{\circ}{=}_{ec} w_{2o} \equiv_{ec} z \text{ (numeral)} \wedge$$

$$\langle t_o(\underline{r}), \ldots \rangle w_{1o} w_{11} \ldots w_{1p} \overset{\circ}{=}_{ec} t_z(\underline{r}) w_{11} \ldots w_{1p} \equiv_{ec} t_z(\underline{r}) w_{21} \ldots w_{2p}$$

$$\equiv_{ec} \langle t_o(r), \ldots \rangle w_{2o} w_{21} \ldots w_{2p} \qquad \text{reduction,(2)}$$

Analogous for $t(\underline{s})$.

Ad $(*_4)$:

$$\{\varphi\} \text{ e-comp} \wedge \bigwedge_i r_i \equiv_{ec} s_i \wedge \underline{w} \text{ e-comp} \Longrightarrow w_o \overset{\circ}{=}_{ec} z \text{ (numeral)} \wedge$$

$$\langle t_o(\underline{r}), \ldots \rangle \underline{w} \overset{\circ}{=}_{ec} t_z(\underline{r}) w_1 \ldots w_p \equiv_{ec} t_z(\underline{s}) w_1 \ldots w_p$$

$$\equiv_{ec} \langle t_o(\underline{s}), \ldots \rangle w_o w_1 \ldots w_p \qquad \text{reduction,(2)}$$

(10.6) $\{\varphi\}$ e-comp (in $\mathcal{F}$) $\Longrightarrow$ t e-comp (in $\mathcal{F}$)

where $\{\varphi\}$ apart from choice sequences are the constants of $t \in \mathcal{F}$.

Proof: (10.5), (10.1)

The first functional domain for which the extensional computability will be shown is $\mathcal{F}_o$ $\equiv$: functional domain of T.

(10.7) $t \varepsilon \mathcal{F}_o \Longrightarrow t$ e-comp in $\mathcal{F}_o$.

Proof: After (10.6) it suffices to consider the functional constants of T(except choice sequences) and to show:

(*) $\bigwedge_{i=o}^{p} r_i \overset{=}{ec} s_i \Longrightarrow \varphi \underline{\underline{r}} \overset{o}{\underset{ec}{=}} \varphi \underline{\underline{s}}$

Case 1: $\varphi \equiv 0,'$ (10.2)

Case 2: φ is generated in T. Proof by induction.

2.1 φ according to (T1),(T2),(T3):
$\varphi \underline{\underline{r}}, \varphi \underline{\underline{s}}$ reduce directly to terms which have at analogous places at most preceding functional constants or r_i resp. s_i. By (10.5) these terms are on account of the (*)- and induction premise extensionally computational equal.

2.2 φ according to (T4):
Because by the (*)-premise $r_o \overset{o}{\underset{ec}{=}} s_o \overset{=}{ec} z$ (numeral), so $\varphi \underline{\underline{r}}, \varphi \underline{\underline{s}}$ reduce to $\varphi z r_2 \ldots r_p$, $\varphi z s_2 \ldots s_p$. Therefore it suffices to prove (*) for the case $r_o \equiv s_o \equiv z$ (numeral) by an inserted induction on z.

2.2.1 $\varphi 0 r_2 \ldots r_p$, $\varphi 0 s_2 \ldots s_p$ reduce directly to $\psi r_2 \ldots r_p$, $\psi s_2 \ldots s_p$. On account of the (*)- and induction hypothesis these terms are extensionally computational equal.

2.2.2 $\varphi z' r_2 \ldots r_p$, $\varphi z' s_2 \ldots s_p$ reduce directly to $\chi(\varphi z) z r_2 \ldots r_p$, $\chi(\varphi z) z s_2 \ldots s_p$. By (*)-premise, induction hypothesis, induction hypothesis of the inserted induction and (10.1),(10.2) these terms are extensionally computational equal.

The next functional domain be $\mathcal{F}_1$ ≡: closure against application of $\mathcal{F}_o$ plus iterated choice sequences over numerals. The proof of (10.7) ($\mathcal{F}_o$ replaced by $\mathcal{F}_1$) gives with (10.6): $t \varepsilon \mathcal{F}_1 \Longrightarrow t$ e-comp in $\mathcal{F}_1$.

Before further functional domains are treated we will convince ourselves that by the extensional computability of the functional domains $\mathcal{F}$ of our interpretation calculi the functional interpretation altogether is constructively settled. For that purpose we define a valuation $\mathfrak{b}_{\mathcal{F}}$ of the T∪BR-formulae• (which are built up from prime formulae $t_1 \overset{o}{=} t_2$ by the propositional connectives) with truth-values w, f.

Definition: Consider A ε T$\cup$BR with constants only in $\mathcal{F}$.

1. A closed prime formula: A $\equiv$ ($t_1 \overset{Q}{=} t_2$) closed

$\mathcal{L}_{\mathcal{F}}(A)=w \Longleftrightarrow: t_1 \overset{Q}{=}_{ec} t_2$, $\mathcal{L}_{\mathcal{F}}(A)=f \Longleftrightarrow$: not $t_1 \overset{Q}{=}_{ec} t_2$

2. A closed:

$\mathcal{L}_{\mathcal{F}}(A)$ is determined by the usual classical w-f-valuation from the $\mathcal{L}_{\mathcal{F}}$-valuations of the prime formulae.

3. A arbitrary:

$\mathcal{L}_{\mathcal{F}}(A)$=:w, if each closed substitution instance of A over $\mathcal{F}$ has the value w; otherwise $\mathcal{L}_{\mathcal{F}}(A)$=:f.

(10.8) K$\supsetneq$T be a subcalculus of T$\cup$BR and $\mathcal{F}$ a functional domain (of the kind considered) which contains the functional domain of K.

Assumption: $\mathcal{F}$ is (relative to itself) extensionally computable.

Assertion: A provable in K $\Longrightarrow \mathcal{L}_{\mathcal{F}}(A)$=w. Especially 0=1 is not provable in K.

Proof: Induction on the length l of a proof of A in K
For simplification we replace (E) after (5.1) by the deductive equivalent schemes:

(U) $\dfrac{A \longrightarrow r \overset{Q}{=} s}{A \longrightarrow t(r) \overset{Q}{=} t(s)}$ (Sy) $u \overset{Q}{=} v \longrightarrow v=u$ (T) $u \overset{Q}{=} v \wedge v=w \longrightarrow u=w$

I. l=0: A is axiom.

(T0): Intuitionistic propositional logic consists only of tautologies. - In a closed substitution instance of $u \overset{Q}{=} u$ the assumed computability gives on both sides the same value.

(Sy),(T): similarly

(T1)-(T3),(T4) 1. line:
In closed substitution instances the left sides reduce immediately to the right sides; consequently the assumed computability ends on both sides with the same value.

(T4) 2. line:
For a closed substitution instance $\varphi(t^{o})'t_1 \ldots t_p \overset{Q}{=} \chi(\varphi t)tt_1 \ldots t_p$ by the assumed extensional computability one gets $t \overset{Q}{=}_{ec} z$ (numeral) and

$\varphi t' t_1 \ldots t_p \overset{Q}{=}_{ec} \varphi z' t_1 \ldots t_p \overset{=}{=}_{ec} \chi(\varphi z) z t_1 \ldots t_p \overset{=}{=}_{ec} \chi(\varphi t) t t_1 \ldots t_p$

premise, reduction,(10.5)

Thus both sides are calculated to the same value.

Lemma: For numerals z_1, z_2 holds:

(1) $-z_1 z_2 \overset{=}{\text{ec}} 0\underbrace{'\cdots'}_{n\text{-times}} \Longleftrightarrow z_2 \equiv z_1\underbrace{'\cdots'}_{n\text{-times}}$ for $n \geq 1$

(2) $-z_1 z_2 \overset{=}{\text{ec}} 0 \Longleftrightarrow z_1$ greater or equal z_2

Thus: $\mathcal{L}_f(z_1 \geq z_2) = w \Longleftrightarrow z_1$ greater or equal z_2

$\mathcal{L}_f(z_1 < z_2) = w \Longleftrightarrow z_1$ less z_2

Proof:

Ad (1): Induction on z_1

I. $-0z_2 \overset{=}{\text{ec}} 0\underbrace{'\cdots'}_{n\text{-times}} \Longleftrightarrow z_2 \overset{=}{\text{ec}} 0\underbrace{'\cdots'}_{n\text{-times}} \Longleftrightarrow z_2 \equiv 0\underbrace{'\cdots'}_{n\text{-times}}$

II. $-z_1' z_2 \overset{=}{\text{ec}} 0\underbrace{'\cdots'}_{n\text{-times}} \Longleftrightarrow \delta(-z_1 z_2) \overset{=}{\text{ec}} 0\underbrace{'\cdots'}_{n\text{-times}} \Longleftrightarrow -z_1 z_2 \overset{=}{\text{ec}} 0\underbrace{'\cdots''}_{(n+1)\text{-times}}$ $(n \geq 1)$

$\Longleftrightarrow z_2 \equiv z_1\underbrace{'\cdots''}_{(n+1)\text{-times}} \equiv (z_1')\underbrace{'\cdots'}_{n\text{-times}}$ (ind.hyp.)

Ad (2): Induction on z_1

I. $-0z_2 \overset{=}{\text{ec}} 0 \Longleftrightarrow z_2 \overset{=}{\text{ec}} 0 \Longleftrightarrow z_2 \equiv 0 \Longleftrightarrow 0$ greater or equal z_2

II. $-z_1' z_2 \overset{=}{\text{ec}} 0 \Longleftrightarrow \delta(-z_1 z_2) \overset{=}{\text{ec}} 0 \Longleftrightarrow -z_1 z_2 \overset{=}{\text{ec}} 0$ or $-z_1 z_2 \overset{=}{\text{ec}} 0'$

$\Longleftrightarrow$ (z_1 greater or equal z_2) or $z_2 \equiv z_1'$ (ind.hyp.,(1))

$\Longleftrightarrow z_1'$ greater or equal z_2

(BR)

Closed substitution formulae for the (BR)-axioms have the form:

$$t_1(\overline{\overline{t_5, -1t_4 * t_6, t_4}}) < t_4^{\circ} \rightarrow \phi t_1 t_2 t_3 t_4^{\circ} t_5 t_6 t_7 \ldots t_p \overset{\circ}{=} t_2 t_7 \ldots t_p$$

$$t_1(\overline{\overline{t_5, -1t_4 * t_6, t_4}}) \geq t_4 \rightarrow \phi t_1 t_2 t_3 t_4 t_5 t_6 t_7 \ldots t_p \overset{\circ}{=} t_3(\phi t_1 t_2 t_3 t_4' (\overline{\overline{t_5, -1t_4 * t_6, t_4}})) t_7 \ldots t_p$$

Here at first the terms of the premises are calculated according to the assumption: $t_4 \overset{\circ}{\underset{\text{ec}}{=}} z_2$, $t_1(\overline{\overline{t_5, -1t_4 * t_6, t_4}}) \overset{\circ}{\underset{\text{ec}}{=}} z_1$ (numerals z_1, z_2). By the above lemma (2) according as z_1 is less resp. greater or equal z_2 the premises are valued by w resp. f. In case of a f-valuation the whole formula has the value w. For a w-valuation we have to show that the terms in the conclusion are computed to the same value. In the first case:

$$\phi t_1..t_4^{\circ}..t_6t_7..t_p \overset{\circ}{=}_{ec} \phi t_1..z_2..t_6t_7..t_p \equiv_{ec} t_2t_7..t_p \quad \text{(reduction,premise)}$$

In the second case:

$$\phi t_1..t_4^{\circ}..t_6t_7..t_p \overset{\circ}{=}_{ec} \phi t_1..z_2..t_6t_7..t_p \equiv_{ec} t_3(\phi t_1t_2t_3z_2'$$

$$\overline{(t_5,-1z_2*t_6,z_2)})t_7..t_p \equiv_{ec} t_3(\phi t_1t_2t_3t_4'(\overline{t_5,-1t_4*t_6,t_4}))t_7..t_p$$

(premise, reduction,(10.5))

<u>II. l≠0:</u> A is conclusion of a rule.

(Mp) $\frac{B,\ B\rightarrow A}{A}$: By induction hypothesis all substitution instances of the premises get the value w and thus also all of the conclusion.

(S) $\frac{A(u^{\alpha})}{A(t^{\alpha})}$ $(u \notin A(0))$: To a substitution into the conclusion corresponds a substitution into the premise. Because the latter has the value w by induction hypothesis, also the conclusion is w-valued.

(U) $\frac{B\rightarrow r\underline{\underline{u}} \overset{\circ}{=} s\underline{\underline{u}}}{B\rightarrow t(r)\overset{\circ}{=}t(s)}$ $(\underline{\underline{u}} \notin B,r,s)$: Consider a closed substitution $*$ of the conclusion $U^* \equiv: (B^* \rightarrow t^*(r^*)\overset{\circ}{=}t^*(s^*))$. By assumption the value of B^* can effectively be determined from within. If it is f so U^* has the value w. But in the case it is w one gets from the premise - which is w-valued by induction hypothesis - $r^*\underline{\underline{u}} \overset{\circ}{=}_{ec} s^*\underline{\underline{u}}$ for all $\underline{\underline{u}}$-substitutions over $\mathcal{F}$ and thus by assumption $r^* \equiv_{ec} s^*$ over $\mathcal{F}$. Therefore by (10.5) and assumption $t^*(r^*) \overset{\circ}{=}_{ec} t^*(s^*)$ holds; so also in this case U^* has value w. Because $*$ was arbitrary thus the conclusion is w-valued.

(CI) $\frac{A(0),\ A(u)\rightarrow A(u')}{A(u^{\circ})}$ $(u \notin A(0))$: $*$ be a closed substitution of the conclusion U which associates to u° the (closed) term t°. By assumption $t \overset{\circ}{=}_{ec} z$ (numeral z); from (10.5) follows by induction on the A-formation: $\mathcal{L}_{\mathcal{F}}(A(u)^*) \equiv \mathcal{L}_{\mathcal{F}}(A^*(t)) = \mathcal{L}_{\mathcal{F}}(A^*(z))$. According to induction hypothesis the following formulae all have value w:

$$A^*(0),\ A^*(0)\rightarrow A^*(0'),\ldots,\ A^*(0'^{\cdots\prime})\rightarrow A^*(\underbrace{0'^{\cdots\prime\prime}}_{\equiv\, z})$$

From this $A^*(z)$ - and thus also U^* - gets the value w. Because $*$ was arbitrary U is w-valued.

(ω) With the same preliminaries as to (CI) each closed substitution instance of the conclusion reduces to one of a premise. So the induction hypothesis yields a w-valuation of the conclusion.

Remark:

The considerations (10.8) can be extended without difficulty to the above described μ-operator and choice sequences (within a suitable term formation perhaps characterized by $\langle r_0, r_1, \ldots \rangle^{\alpha 0} u^0 = r_u$).

For closed formulae the assertion of (10.8) can be inverted under the above made assumptions:

$$A \text{ closed} \wedge \mathcal{W}_{\mathcal{F}}(A)=w \Longrightarrow A \text{ provable in } K$$

This simply follows from the fact that closed prime formulae are decidable in K by pursueing the computation of both sides in K.

We are now in the position to give a full survey of our functional interpretation.

(10.9) Let $K \supseteq T$ be a subcalculus of $T \cup BR$ and $\mathcal{F}$ a functional domain which contains the functional domain of K.

Assumption: $\mathcal{F}$ is (relative to itself) extensionally computable.

Then the following holds: If a classical (enumerable) theory $\mathcal{T}$ (over the functional language) is functional interpretable in K, so $\mathcal{T}$ is in this way constructively interpreted in K; especially then $\mathcal{T}$ is ω-consistent and only such functionals can be proved in $\wedge\vee$-form to exist in $\mathcal{T}$ which are in K. If moreover the "quantifierfree" axiom of choice (AC)-qf is in $\mathcal{T}$ then to each formula provable in $\mathcal{T}$ a in $\mathcal{T}$ classical equivalent $\wedge\vee$-formula is directly constructively interpreted in K. - Analogous statements hold for the functional interpretation in the narrower sense of an intuitionistically approximated theory $\mathcal{T}^*$.

We briefly summarize this in saying that <u>the functional interpretation (in the narrower sense) of $\mathcal{T}$ ($\mathcal{T}^*$) is constructively given in K.</u>

Proof: (4.5),(10.8); (4.7): extension of $\mathcal{F}$ by the e-comp functionals (+) of the premise (b) of (4.7).

(10.10) The functional interpretation of classical number theory and the functional interpretation in the narrower sense of Heyting-analysis plus (MP), ($\underrightarrow{V}$) is constructively given in T.
Proof: (6.3),(6.4),(10.7),(10.9)

Now we continue the investigation of the functional domains. Next we consider $\mathcal{F}_2$ ≡: closure against applications of T plus iterated choice sequences over numerals, "fixed choice sequences" and μ. The terms of $\mathcal{F}_2$ can be generated as follows:

1. Generation processes of T accordingly extended to $\mathcal{F}_2$.
2. $\mu \varepsilon \mathcal{F}_2$
3. Iterated choice sequences over numerals are in $\mathcal{F}_2$.
4. $r^{\alpha o}, r_o^{\alpha}, \dots, r_z^{\alpha}, s^{\alpha} \ \varepsilon \ \mathcal{F}_2 \implies \langle r0, r1, \dots \rangle, \langle r_o, \dots, r_z, r0, r1, \dots \rangle,$
$\langle r_o, \dots, r_z, s, s, \dots \rangle \ \varepsilon \ \mathcal{F}_2$ ("fixed choice sequences")
5. $r^{\alpha\beta}, s^{\alpha} \ \varepsilon \ \mathcal{F}_2 \implies (rs)^{\alpha} \ \varepsilon \ \mathcal{F}_2$

Without iterated choice sequences over numerals this gives an effective system of notations.

The μ-computability is based on the following general continuity theorem.

(10.11) Consider a functional domain $\mathcal{F}$.
Of the choice sequences in a computable term $t^o \varepsilon \mathcal{F}$ the evaluation uses (only finitely many and of these) only a finite initial segment, i.e. with $t \equiv t(\dots\langle r_{io}, r_{i1}, \dots\rangle\dots)$ (This notation refers to all choice sequences in t.) holds:

$$t^o \text{ e-comp in } \mathcal{F} \implies \bigvee \text{numeral } z \bigwedge i, \bigwedge s_{io}, s_{i1}, \dots \varepsilon \mathcal{F} \ t(\dots\langle r_{io}, \dots\rangle\dots) \overset{o}{=}_{ec} t(\dots\langle r_{io}, \dots, r_{iz}, s_{io}, s_{i1}, \dots\rangle\dots)$$

Denotation: z continuity point

Remark:

1. The reference to choice sequences and not to arbitrary subterms of type $\alpha 0$ is at this stage of the argumentation necessary. For instance with φ according to (T2), $\psi \equiv 0$ the term φt^{o} always reduces to 0 independently of t. But for $<\varphi 0,\varphi 1,\ldots>^{oo}t^{o}$ one first has to compute t^{o}. As long as a total computability is not certain this generalization cannot be stated.

2. In the proof of (10.11) one has to consider firstly full choice sequences. Only following up (10.11) one can restrict oneself to sufficient long finite choice sequences $<r_o,\ldots,r_z>$. In presence of extensionality the latter may still be replaced by a T-functional over $r_o,\ldots,r_z$.

Proof: Induction on the length l of a computation of t^{o}, i.e. the number of nodes in the computation tree of t.

I. l=1: t is numeral. Assertion trivial.

II. l>1: Distinction between cases by the first reduction step, i.e. by the first functional sign φ in $t^{o} \equiv \varphi t_1 \ldots t_p$. Choice sequences appear in t at most in $t_1,\ldots,t_p$ respectively φ itself may be a choice sequence. Now if one looks at the reduction instructions then it follows immediately that in all reduction steps with the exception of the case that φ is choice sequence and t_1^{o} numeral all choice sequences - if at all - pass over completely unchanged into the direct subterms. In these cases the maximum of the continuity points of these finitely many subterms which exist by induction hypothesis is a continuity point of t. - In the above mentioned rest case φ choice sequence, t_1^{o} numeral, only the choice sequence φ is dissolved while all other possibly occurring choice sequences pass over unchanged. Here the maximum of t_1^{o} and the continuity point of the subterm which exists by induction hypothesis is a continuity point of t.

(10.12) $t \varepsilon \mathcal{F}_2 \Longrightarrow$ t e-comp in $\mathcal{F}_2$

Proof: We only have to continue the proof of (10.7) ($\mathcal{F}_o$ replaced by $\mathcal{F}_2$); because of (10.6) to the cases already considered only μ is added.

2.3 $\varphi \equiv \mu$: We have to show $\bigwedge_{i=1}^{2} r_i \overset{=}{_{ec}} t_i \Longrightarrow \mu r_1 r_2 \overset{o}{\underset{ec}{=}} \mu t_1 t_2$

$\bigwedge_{i=1}^{2} r_i =_{ec} t_i \Longrightarrow r_1, r_2, t_1, t_2$ e-comp $\wedge$ $t_1(<t_2 0, \ldots>)$ e-comp $\wedge$
$\bigvee$numeral $\hat{z} \bigwedge s_0, s_1, \ldots \varepsilon \mathcal{F}, \bigwedge i \; t_1(<t_2 0, \ldots>) \overset{o}{=}_{ec}$
$t_1(<t_2 0, \ldots, t_2 \hat{z}, s_0, \ldots, s_i>) \wedge \bigwedge$numerals $i \geq 1$
$t_1(<t_2 0, \ldots, t_2(i-1)>) \overset{o}{=}_{ec} s_1(<s_2 0, \ldots, s_2(i-1)>)$
(10.5),(10.2),(10.1);(10.11)

$\Longrightarrow$ Analogous subterms of $\mu r_1 r_2$, $\mu t_1 t_2$ have the same value; because of the continuity these two rows of subterms therefore break off at the same point; $\mu r_1 r_2 \overset{o}{=}_{ec} \mu t_1 t_2$.

Remark:
Analogously further μ-formations e.g. $\mu x^0(w(<<v00, \ldots, v0(x-1)>, \ldots, <v(x-1)0, \ldots, v(x-1)(x-1)>>)<x)$, can be treated.

The last functional domain of this chapter be $\mathcal{F}_3 \equiv$: closure against applications of T plus iterated choice sequences over numerals, "fixed choice sequences", μ and BR^{τ_i} $(i=0,1,\ldots)$ = closure against applications of $\mathcal{F}_2$ plus BR^{τ_i} $(i=0,1,\ldots)$, where $\tau_0 \equiv: 0$, $\tau_{n+1} \equiv: \tau_n 0$. The $\mathcal{F}_3$-terms can be generated analogously to that of $\mathcal{F}_2$ if the bar recursive functionals of the types 0, 00, 000, ... are added.

(10.13) $\bigwedge$nat i, $\bigwedge$e-comp t^{τ_i} $\bigvee$iterated choice sequence α^{τ_i} over numerals:
$t =_{ec} \alpha$

<u>Denotation:</u>
From now on "cs", "ics" will stand for "choice sequence" resp. "iterated choice sequence over numerals"; herein the numerals as lowest level are included.

<u>Proof:</u> Induction on i

I. i=0: $\bigwedge$e-comp $t^0 \bigvee$numeral z $t \overset{o}{=}_{ec} z$ according to definition

II. Induction hypothesis: $\bigwedge$e-comp $s^{\tau_i} \bigvee$ics $\alpha_s^{\tau_i}$ $s =_{ec} \alpha_s$

$t^{\tau_{i+1}}$ e-comp $\Longrightarrow \bigwedge$ numerals $j\{(tj)^{\tau_i}$ e-comp $\wedge$ $tj =_{ec} \alpha_{tj}\}$ (10.2),(10.1), ind.hyp.

$\Longrightarrow s_1 \overset{o}{=}_{ec} s_2 \Longrightarrow s_1 =_{ec} s_2 =_{ec} z$ (numeral) $\wedge$ $ts_1 =_{ec} tz =_{ec} \alpha_{tz}$

$=_{ec} <\alpha_{t0}, \ldots>z =_{ec} <\alpha_{t0}, \ldots>s_2$ reduction

$t^{\tau_{i+1}}$ e-comp $\Longrightarrow t \overset{=}{\underset{ec}{}} \langle \alpha_{to}, \alpha_{t1}, \ldots \rangle$

(10.14) $t \varepsilon \mathcal{F}_3 \Longrightarrow t$ e-comp in $\mathcal{F}_3$

<u>Proof:</u> We have to complete the proofs of (10.7),(10.12) ($\mathcal{F}_o$, $\mathcal{F}_2$ replaced by $\mathcal{F}_3$) by the bar recursive functionals $\phi^{(\tau_i)}$ (i=0,1,...).

2.4 $\varphi \equiv \phi^{(\tau_i)}$:

It is to be shown:

$$(*_1) \quad \bigwedge_{i=1}^{p} r_i \overset{=}{\underset{ec}{}} t_i \Longrightarrow \phi r_1 r_2 r_3 r_4^{o} r_5^{\tau_i o} r_6^{\tau_i} \ldots r_p \overset{o}{\underset{ec}{}} \phi t_1 t_2 t_3 t_4 t_5 t_6 \ldots t_p \quad (p \geq 6)$$

Because of $r_4 \overset{o}{\underset{ec}{}} t_4 \overset{=}{\underset{ec}{}} z$ (numeral) $(*_1)$ reduces immediately to the case $r_4 \equiv t_4 \equiv z$ (numeral):

$$(*_2) \quad \bigwedge_{\substack{i=1\\ i\neq 4}}^{p} r_i \overset{=}{\underset{ec}{}} t_i \Longrightarrow \phi r_1 r_2 r_3 z r_5^{\tau_i o} r_6^{\tau_i} \ldots r_p \overset{o}{\underset{ec}{}} \phi t_1 t_2 t_3 z t_5 t_6 \ldots t_p$$

for all numerals z

From the definitions and preceding considerations (10.5), (10.7), lemma in (10.8), page 106 follows:

(1) $\overline{\overline{v,-1u*w,u}}$ is extensionally computable in the arguments u,v,w.

(2) $\overline{\overline{v,-1u*w,u}} \overset{=}{\underset{ec}{}} \langle v0, \ldots, v(u-2), w \rangle$

↑ position u-1

Here we have the agreement that negative positions don't exist.

By $(*_2)$-premise and (1) $r_1(\overline{\overline{r_5,-1z*r_6,z}}) \overset{o}{\underset{ec}{}} t_1(\overline{\overline{t_5,-1z*t_6,z}})$ holds; the first subterms s_1^o in the reduction of both sides of the $(*_2)$-conclusion have the same value so that the decision for the second subterms s_o^o is analogous on both sides.

<u>Case 1:</u> value/$s_1 < z$

$$\phi r_1 r_2 r_3 z r_5 r_6 \ldots r_p \overset{o}{\underset{ec}{}} r_2 r_7 \ldots r_p \overset{=}{\underset{ec}{}} t_2 t_7 \ldots t_p \overset{=}{\underset{ec}{}} \phi t_1 t_2 t_3 z t_4 t_5 \ldots t_p$$

reduction, $(*_2)$-premise, (10.5)

<u>Case 2:</u> value/$s_1 \geq z$

According to the reduction:

$$(3) \quad \begin{cases} \phi r_1 r_2 r_3 z r_5 r_6 \ldots r_p \overset{o}{\underset{ec}{}} r_3(\phi r_1 r_2 r_3 z'(\overline{\overline{r_5,-1z*r_6,z}}))r_7 \ldots r_p \\ \phi t_1 t_2 t_3 z t_5 t_6 \ldots t_p \overset{=}{\underset{ec}{}} t_3(\phi t_1 t_2 t_3 z'(\overline{\overline{t_5,-1z*t_6,z}}))t_7 \ldots t_p \end{cases}$$

Now it suffices to show:

$$(*_3) \quad \bigwedge_{i=1}^{3} t_i \overset{=}{ec} r_i \wedge r_5 \overset{\tau_i o}{\underset{ec}{=}} t_5 \overset{=}{ec} \langle s_o,\dots,s_{z-1}\rangle \Longrightarrow$$

$$\phi r_1 r_2 r_3 z' r_5 \overset{=}{ec} \phi t_1 t_2 t_3 z' t_5 \qquad \text{for numerals } z$$

Because from this by (3), ($*_2$)-premise, (1), (2), (10.5) the ($*_2$)-conclusion follows at once.

For ($*_3$) we now set up a bar induction over the "spread of the iterated choice sequences of type τ_i". Because the $\bigwedge$-statement in the sixth argument of the ($*_3$)-conclusion terms must be taken over into the "induction predicate" Q, for each node we have to consider all ramifications. Therefore the effective "delimination predicate" P has to refer to the continuity points. This motivates the following definitions.

Given $\bigwedge_{i=1}^{3} t_i \overset{=}{ec} r_i \wedge s_o,\dots,s_{z-1}$ e-comp.

$P(f^{\tau_{i+1}},u^o) \equiv: \bigvee$numeral $y<u\{$computation of $r_1(\langle s_o,\dots,s_{z-1},f0,..,f(y-1)\rangle)^o$ uses of the denoted argument $\langle s_o,\dots,s_{z-1},f0,..,f(y-1)\rangle$ at most the first $z+y$ links and gives a value $<z'+y-1\}$

$Q(f^{\tau_{i+1}},u^o) \equiv: P(f,u) \vee \bigwedge \underline{r},\underline{t}\{ \bigwedge_{i=6}^{p} r_i \overset{=}{ec} t_i \wedge r_5 \overset{=}{ec} t_5 \overset{=}{ec} \langle s_o,..,s_{z-1},f0,..,f(u-1)\rangle \Longrightarrow \phi r_1 r_2 r_3 (z'+u) r_5 r_6..r_p \overset{o}{\underset{ec}{=}} \phi t_1 t_2 t_3 (z'+u) t_5 t_6..t_p\}$

for ics $f^{\tau_{i+1}}$ and numerals u^o. ($r_1(\langle s_o,..,s_{z-1},f0,..,f(y-1)\rangle)$ is e-comp on account of the made assumptions.)

Immediately follows:

(4) $P(f,u) \vee \neg P(f,u)$; $P(f,u)$ is decidable.

(5) $P(f,u) \longrightarrow P(f,u')$; $P(f,u) \longrightarrow Q(f,u)$; $\neg P(f,0)$

Because in P, Q the $\mathcal{F}_3$-prime formulae are $\underset{ec}{=}$-extensional in all denoted parts and the computation is only determined by the first functional sign of a term, so the in general intensional P - and thus also Q - is at least with respect to choice sequences whose links are reducible to each other extensional.

(6) P, Q "choice sequence"-extensional in f,u. From this one gets at once:

(7) $P(f,u) \longleftrightarrow P(\langle f0,\dots,f(u-1)\rangle,u) \longleftrightarrow P(\langle f0,\dots,f(u-2)\rangle,u)$

(8) $Q(f,u) \longleftrightarrow Q(\langle f0,\ldots,f(u-1)\rangle,u)$

(9) $\bigwedge ics\ f^{\tau_{i+1}} \bigvee numeral\ x^{o}\ P(\langle f0,\ldots,f(x-1)\rangle,x)$

<u>Proof:</u> From the assumptions follows the extensional computability of $r_1(\langle s_o,\ldots,s_{z-1},f0,f1,\ldots\rangle)^{o}$. After (10.11) one gets to a continuity point which can be exceeded so far that also the value statement for P is satisfied: $\bigvee x^{o} P(f,x)$. (9) now by (7).

(10) $\bigwedge ics\ g^{\tau_i}\ Q(\langle f0,\ldots,f(u-1),g\rangle,u') \Longrightarrow Q(\langle f0,\ldots,f(u-1)\rangle,u)$

<u>Proof:</u> Because of (4) the following distinctions by cases are allowed.

<u>Case 1:</u> $P(\langle f0,\ldots,f(u-1)\rangle,u')$

<u>Case 1.1:</u> $P(\langle f0,\ldots,f(u-1)\rangle,u)$
Hence $Q(\langle f0,\ldots,f(u-1)\rangle,u)$ with (5).

<u>Case 1.2:</u> $\neg P(\langle f0,\ldots,f(u-1)\rangle,u)$
Thus the computation of $r_1(\langle s_o,\ldots,s_{z-1},f0,\ldots,f(u-1)\rangle)^{o}$ uses of the $\langle\ldots\rangle$-argument at most the first z+u links and gives a value $\langle z'+u-1 < z'+u$. Since from these premises for e-comp g $r_1(\langle s_o,\ldots,s_{z-1},f0,\ldots,f(u-1),g\rangle) \overset{\circ}{=}_{ec} t_1(\langle s_o,\ldots,s_{z-1},f0,\ldots,f(u-1),g\rangle) < z'+u$ follows, so now one gets by (1),(2) on account of the existing extensionality:

$$\bigwedge_{i=6}^{p} r_i =_{ec} t_i \wedge r_5 =_{ec} t_5 =_{ec} \langle s_o,\ldots,s_{z-1},f0,\ldots,f(u-1)\rangle \Longrightarrow$$
$$\phi t_1 t_2 t_3 (z'+u) t_5 t_6 \ldots t_p \overset{\circ}{=}_{ec} t_2 t_7 \ldots t_p =_{ec} r_2 r_7 \ldots r_p =_{ec}$$
$$\phi r_1 r_2 r_3 (z'+u) r_5 r_6 \ldots r_p \qquad \text{premise, reduction}$$

Thus $Q(f,u)$ and by (8) $Q(\langle f0,\ldots,f(u-1)\rangle,u)$.

<u>Case 2:</u> $\neg P(\langle f0,\ldots,f(u-1)\rangle,u')$
$\neg P(\langle f0,\ldots,f(u-1)\rangle,u)$ (5)

The implication hypothesis of (10) says:

$$\bigwedge ics\ g^{\tau_i} (P(\langle f0,\ldots,f(u-1),g\rangle,u') \vee \bigwedge \underline{\underline{r}},\underline{\underline{t}} \{ \bigwedge_{i=6}^{p} r_i =_{ec} t_i \wedge r_5 =_{ec} t_5 =_{ec}$$
$$\langle s_o,\ldots,s_{z-1},f0,\ldots,f(u-1),g\rangle \Longrightarrow \phi r_1 r_2 r_3 (z'+u') r_5 r_6 \ldots r_p \overset{\circ}{=}_{ec}$$
$$\phi t_1 t_2 t_3 (z'+u') t_5 t_6 \ldots t_p \})$$

As now $P(\langle f0,\ldots,f(u-1),g\rangle,u') \longleftrightarrow P(\langle f0,\ldots,f(u-1)\rangle,u') \longleftrightarrow \bot$ (7), case so by (10.1) this means:

$$(*)\ \bigwedge ics\ g^{\tau_i} \bigwedge r_5,t_5 \{ r_5 =_{ec} t_5 =_{ec} \langle s_o,\ldots,s_{z-1},f0,\ldots,f(u-1),g\rangle \Longrightarrow$$
$$\phi r_1 r_2 r_3 (z'+u)' r_5 =_{ec} \phi t_1 t_2 t_3 (z'+u)' t_5 \}$$

Assume $\bigwedge_{i=6}^{p} r_i \overset{=}{\text{ec}} t_i$, $r_5 \overset{=}{\text{ec}} t_5 \overset{=}{\text{ec}} \langle s_o,\ldots,s_{z-1},f0,\ldots,f(u-1)\rangle$.

<u>Case 2.1:</u> $\text{value}/r_1(\langle s_o,\ldots,s_{z-1},f0,\ldots,f(u-1),t_6\rangle)<z'+u$

With the premises and (1),(2) the direct reduction yields:

$$\phi r_1 r_2 r_3 (z'+u) r_5 r_6 \ldots r_p \overset{\text{o}}{\underset{=}{\text{ec}}} r_2 r_7 \ldots r_p \overset{=}{\text{ec}} t_2 t_7 \ldots t_p \overset{=}{\text{ec}}$$
$$\phi t_1 t_2 t_3 (z'+u) t_5 t_6 \ldots t_p .$$

<u>Case 2.2:</u> $\text{value}/r_1(\langle s_o,\ldots,s_{z-1},f0,\ldots,f(u-1),t_6\rangle)\geq z'+u$

With the premises, (1), (2) the direct reduction now gives:

$$(**)\begin{cases} \phi r_1 r_2 r_3 (z'+u) r_5 r_6 \ldots r_p \overset{\text{o}}{\underset{=}{\text{ec}}} \\ \qquad r_3(\phi r_1 r_2 r_3 (z'+u)'(\overline{\overline{r_5,-1(z'+u)*r_6,z'+u}}))r_7\ldots r_p \\ \phi t_1 t_2 t_3 (z'+u) t_5 t_6 \ldots t_p \overset{\text{o}}{\underset{=}{\text{ec}}} \\ \qquad t_3(\phi t_1 t_2 t_3 (z'+u)'(\overline{\overline{t_5,-1(z'+u)*t_6,z'+u}}))t_7\ldots t_p \end{cases}$$

After (10.13) for the e-comp $r_6^{\tau_i}$ there is an ics g^{τ_i} with: $t_6 \overset{=}{\text{ec}} r_6 \overset{=}{\text{ec}} g$.

Together with the r_5,t_5-premise, (1), (2) this yields

$$\overline{\overline{r_5,-1(z'+u)*r_6,z'+u}} \overset{=}{\text{ec}} \langle s_o,\ldots,s_{z-1},f0,\ldots,f(u-1),g\rangle$$
$$\overset{=}{\text{ec}} \overline{\overline{t_5,-1(z'+u)*t_6,z'+u}}$$

Therefore one gets by (*)

$$\phi r_1 r_2 r_3 (z'+u)'(\overline{\overline{r_5,-1(z'+u)*r_6,z'+u}}) \overset{=}{\text{ec}}$$
$$\phi t_1 t_2 t_3 (z'+u)'(\overline{\overline{t_5,-1(z'+u)*t_6,z'+u}})$$

and with the premises, (**), (10.1)

$$\phi r_1 r_2 r_3 (z'+u) r_5 r_6 \ldots r_p \overset{\text{o}}{\underset{=}{\text{ec}}} \phi t_1 t_2 t_3 (z'+u) t_5 t_6 \ldots t_p .$$

Thus both subcases hold Q(f,u), i.e. after (8) Q(<f0,...,f(u-1)>,u).

From (4) - (10) follows by bar induction $BI_D^{\tau_i}$ over ics of type τ_i applied to P,Q: $Q(\sigma,0)$, i.e. by (10.1)

$$P(\sigma,0) \vee \bigwedge r_5,t_5\{r_5 \overset{=}{\text{ec}} t_5 \overset{=}{\text{ec}} \langle s_o,\ldots,s_{z-1}\rangle \Longrightarrow$$
$$\phi r_1 r_2 r_3 z' r_5 \overset{=}{\text{ec}} \phi t_1 t_2 t_3 z' t_5\}$$

On account of (5) thus $(*_3)$ is proved.

Because up to now BI_D is intuitionistically accepted only for type 0 this proof gives for the present only a foundation of $(BR)^o$. But by Howard-Kreisel [19], §7 under certain conditions bar induction for the types $\tau_i \equiv 0\ldots0$ can be reduced to Brouwer's bar induction BI_D^o. The here interesting part reads as follows:

<u>Theorem:</u> Intuitionistically for "choice sequence"-extensional predicates $P(f^{\sigma oo},u^o)$, $Q(f^{\sigma oo},u^o)$ bar induction $BI_D^{\sigma o}$ follows from BI_D^{σ} and the following strong continuity for P:

$(SC)^{\sigma}$ $\bigwedge cs\ f^{\sigma o} \bigvee x^o \bigvee y^o{<}x\ P({<}f_o^{\sigma o},\ldots,f_{y-1}{>},y) \Longrightarrow$

$\bigvee functional\ g^{o(\sigma o)o}$ over $\sigma 0$-$cs(\bigwedge cs\ f^{\sigma o} \bigvee n^o\ g({<}f0,\ldots,f(n-1){>})n \neq 0\ \wedge$

$\bigwedge cs\ f^{\sigma o} \bigwedge h^{\sigma},n^o\{g({<}f0,..,f(n-1){>})n \neq 0 \Longrightarrow g({<}f0,..,f(n-1),h{>})n' \overset{o}{=}$

$g({<}f0,..,f(n-1){>})n \neq 0\} \wedge \bigwedge cs\ f^{\sigma o} \bigwedge n^o\{g({<}f0,\ldots,f(n-1){>})n \neq 0 \Longrightarrow$

$\bigvee y^o{<}g({<}f0,\ldots,f(n-1){>})n - 1\ \ P({<}f_o,\ldots,f_{y-1}{>},y)\})$

where the cs $f_n^{\sigma o} =: \lambda y^o f^{\sigma o}({<}{<}n,y{>}{>})$, ${<}{<},{>}{>}$ a primitive recursive one-to-one pairing function from $N \times N$ onto N.

<u>Proof:</u> Theorem 7A,7B in [19], §7; lemma 7.1 contains the here assumed extensionality; the proof is in the there used [..]-translation of the iterated choice sequences noted only for $\sigma \equiv 0$, but it can be read for arbitrary σ in our language likewise; gc corresponds here to gc(lh c). - Observe that no reference to a "higher" spread is necessary!

If herein σ is replaced by τ_o and "cs" by "ics" so the connections are not changed and one gets intuitionistically with (6) from Brouwer's bar induction BI_D^o the above used bar induction case $BI_D^{\tau_1}$ because $(SC)^{\tau_o}$ over ics can be established as follows.

To begin with from the P-definition,(6) follows:

(11) $\bigvee u^o{<}x\ P({<}f_o,..,f_{u-1}{>},u) \Longleftrightarrow \bigvee y^o{<}x-1\{$computation of
$r_1({<}s_o,\ldots,s_{z-1},f_o,\ldots,f_{y-1}{>})^o$ uses of the
${<}\ldots{>}$-argument at most the first z+y links and
gives a value ${<}z'+y-1\}$

For ics $f^{\tau_o^o}$ (10.11) gives to the e-comp $r_1({<}s_o,..,s_{z-1},{<}f_o0,..{>},{<}f_10,..{>},\ldots{>})^o$ effectively a continuity point l, such that for all extensions of the choice sequences the value/$r_1({<}s_o,..,s_{z-1},{<}f_o0,....,f_o(l-1){>},\ldots,{<}f_{l-1}0,..f_{l-1}(l-1){>}{>})$ is unchanged and the computation uses at most the denoted links of the choice sequences.

$g_1 f$ =: this 1 for ics $f^{\tau_0 0}$

$g_2 f$ =: $\max(g_1 f, \text{value}/r_1(<s_0,..,s_{z-1},<f_0 0,..,f_0(g_1 f-1)>,...$
$<f_{g_1 f-1} 0,..,f_{g_1 f-1}(g_1 f-1)>>))$ for ics $f^{\tau_0 0}$

By this $\forall y^0 < g_2 f+2$ {computation of $r_1(<s_0,..,s_{z-1},f_0,..,f_{y-1}>)$ uses of the <..>-argument at most the first z+y links and gives a value <z'+y-1 ; thus with (11)

(12) $\forall u^0 < g_2 f+3 \ P(<f_0,...,f_{u-1}>,u)$

The premise of $(SC)^{\tau_0}$ is satisfied; we have to show the conclusion. - But the calculation of $g_1 f$, $g_2 f$ requires by the foregoing only finitely many $f(<<a^0,b^0>>)$-values, namely only such with $a,b<g_1 f$.

$g_3 f$ =: $\max\{<<a,b>> \mid a,b<g_1 f\}$

therefore holds, due to extensionality

(13) $g_3 f = g_3(<f0,...,f(g_3 f),g0,g1,...>)$
(14) $g_2 f = g_2(<f0,...,f(g_3 f),g0,g1,...>)$ for arbitrary ics $g^{\tau_0 0}$

(15) $gfn \overset{o}{=}: \begin{cases} g_2 f+4 & \text{if } n>g_3 f \\ 0 & \text{otherwise} \end{cases}$ for ics $f^{\tau_0 0}$

This g fulfills the $(SC)^{\tau_0}$-conclusion, for:

$g(<f0,..,f(g_3 f)>)(g_3 f+1) = g(<f0,..,f(g_3 f)>)(g_3(<f0,..,f(g_3 f)>)+1) \neq 0$
(13),(15)

$g(<f0,..,f(n-1)>)n \neq 0 \Longrightarrow n>g_3(<f0,..,f(n-1)>)=g_3(<f0,..,f(n-1),h>)$
$=g_3 f \wedge g_2(<f0,..,f(n-1)>)=g_2(<f0,..,f(n-1),h>)$
$=g_2 f$ (15),(13),(14)
$\Longrightarrow g(<f0,..,f(n-1),h>)n'=g(<f0,..,f(n-1)>)n$
$=gfn \neq 0 \wedge \forall u<gfn-1=g_2 f+3 \ P(<f_0,..,f_{u-1}>,u)$
(12),(15)

Quite analogous to this reduction of $BI_D^{\tau_1}$ over ics to BI_D^0 by $(SC)^{\tau_0}$ in general the reduction of $BI_D^{\tau_i}$ over ics to BI_D^0 by $(SC)^{\tau_{i-1}}$ comes off. Now one considers in the Howard-Kreisel theorem instead of 2-tupels (i+1)-tupels of numbers and in the proof of $(SC)^{\tau_{i-1}}$ the correspondingly manyfold nested iterated finite choice sequences over numerals.

With this the proof of (10.14) is finished.

In the argumentation for bar recursion choice sequences are used because only with their help in applying (10.11) we were able to dispose of the bar for the bar recursion process effectively. From the so obtained knowledge (10.11), (10.14) now the extensional computability of the corresponding T$\cup$BR-part (relative to itself) can be inferred.

(10.15) Let $\mathcal{F}$ be a functional domain which belongs to a subcalculus K$\supseteq$T of T$\cup$BR and let $\hat{\mathcal{F}} \supseteq \mathcal{F}$ be a relative to itself extensionally computable functional domain which is closed against $\mathcal{F}$-choice sequences over the types of the bar recursive functionals appearing in $\mathcal{F}$.

Then also $\mathcal{F}$ is extensionally computable relative to itself.

<u>Proof:</u> Because K is a subcalculus so $\mathcal{F}$ is effectively delimitable and the computation of $\mathcal{F}$ takes its course entirely in $\mathcal{F}$. Therefore we may again follow along the previous considerations. We proceed from (10.7) and treat the $\mathcal{F}_o$-part of $\mathcal{F}$ analogously. The bar recursive functionals of $\mathcal{F}$ are dealed with in analogy to (10.14), page 112-115 under the following modifications. All parts containing choice sequences are referred to $\hat{\mathcal{F}}$; all other parts refer to $\mathcal{F}$. One reads f as a choice sequence over the effectively denumerable $\mathcal{F}$-functionals of the type considered. fi denotes the $\mathcal{F}$-object at the i-th position in f. The considerations for this come off almost parallel to (10.14); (6), (7), (8) follow again immediately. Whereas also (9) carries over unchanged for (10) one has to use the fact that under the assumptions of the theorem and on account of (10.7) for

r_1 $\mathcal{F}$-e-comp, $r_5 \underset{\mathcal{F}}{\overset{=}{\mathrm{ec}}} t_5 \underset{\hat{\mathcal{F}}}{\overset{=}{\mathrm{ec}}} \langle s_o,..,s_{z-1},f0,..,f(u-1)\rangle$, $r_6 \underset{\mathcal{F}}{\overset{=}{\mathrm{ec}}} t_6$ now

$$\overline{\overline{r_5,-1(z'+u)*r_6,z'+u}} \underset{\mathcal{F}}{\overset{=}{\mathrm{ec}}} \overline{\overline{t_5,-1(z'+u)*t_6,z'+u}}$$

$$\underset{\hat{\mathcal{F}}}{\overset{=}{\mathrm{ec}}} \langle s_o,..,s_{z-1},f0,..,f(u-1),t_6\rangle \quad \text{and}$$

$$r_1(\overline{\overline{r_5,-1(z'+u)*r_6,z'+u}}) \underset{\mathcal{F}}{\overset{\overset{\circ}{=}}{\mathrm{ec}}} t_1(\overline{\overline{t_5,-1(z'+u)*t_6,z'+u}})$$

$$\underset{\hat{\mathcal{F}}}{\overset{=}{\mathrm{ec}}} r_1(\langle s_o,..,s_{z-1},f0,..,f(u-1),t_6\rangle)$$

holds. So with Brouwer's bar induction over the spread of the effectively denumerable $\mathcal{F}$-functionals of the type considered, one gets the extensional computability of the BR-functionals of $\mathcal{F}$.

(10.15) applied to the T$\cup$BR-part of f_3 gives from (10.14):

(10.16) $T\cup BR^{0\ldots 0}$ is extensionally computable relative to itself.

Finally with (10.9), (8.2), (9.1) we have the following total result.

(10.17) The functional interpretation of that part of classical analysis which is based on $(AC)^{0,0\ldots 0}$-Λ, $(\omega AC)^{0\ldots 0}$-Λ, $(C)^{0}$-$\Delta^{0\ldots 0,\beta}$ is constructively (intuitionistically) given in $T\cup BR^{0\ldots 0}$. This includes arithmetical and hyper-arithmetical comprehension; however after Kreisel [27], 151 these means don't cover Π_1^1-analysis or $(C)^0$-Δ_2^1 for which after (8.2), (9.1) bar recursion of type 0(00) is necessary. - Corresponding statements hold for the functional interpretation in the narrower sense of the previously treated intuitionistic approximated theories.

The above given foundation of bar recursion could be carried over to higher types provided that we dispose of functional domains which are at least in an extensional representation closed against choice sequences over these types. Previously this was simply achieved by the fact that ics over numerals cover extensionally all possibilities for the types 0...0. For higher types a survey is not so easy and it has particularly to be proven that choice sequences over a type of the functional domain together with the other operations yield no new elements. - In the following after the idea of Gödel using generalized inductive definitions we try to build up as intuitionistically as possible sufficient large constructive functional domains (models) which on one side are closed against T$\cup$BR and choice sequences in the types considered and on the other side are not too large so that for their existence and the further considerations there still will be a certain intuitionistic evidence - mainly in the sense of an extension of hitherto formulated intuitionistic principles.

XI. Generalized inductive definitions

Firstly we shall adopt the classical standpoint. Behind generalized inductive definitions here we find the theory of chains founded by Dedekind and further developed by Zermelo and Hessenberg [13]. We shall give a short survey and take as a basis a typed domain of sets (in analogy to the avoidance of the paradoxes in the functional treatment).

Let $\mathfrak{A}$ be a predicate over things which is in the bounds U, O monotone in sets:

(0) $U \subseteq O$

(1) $U \subseteq P_1 \subseteq P_2 \subseteq O \rightarrow U \subseteq \mathfrak{A}P_1 \subseteq \mathfrak{A}P_2 \subseteq O$

For this the generalized inductive definition asserts the existence of a set $\bar{U}$ with:

(2) $\mathfrak{A}\bar{U} \subseteq \bar{U}$ (3) $\mathfrak{A}Q \subseteq Q \wedge U \subseteq Q \subseteq O \rightarrow \bar{U} \subseteq Q$ (4) $U \subseteq \bar{U} \subseteq O$

This can be proved by full comprehension (C) but otherwise intuitionistically as follows.

(5) $\mathfrak{K}P \equiv: \mathfrak{A}P \subseteq P$ (chain from above)

(6) $\bar{U}a \leftrightarrow: \bigwedge P(U \subseteq P \subseteq O \wedge \mathfrak{K}P \rightarrow Pa)$ (intersection of <u>all</u> chains from above between U,O) (C)

From (6) follows at once (3): $\mathfrak{K}P \wedge U \subseteq P \subseteq O \rightarrow \bar{U} \subseteq P$. Because by (1) $\mathfrak{K}O$ holds, so (6) also gives immediately (4): $U \subseteq \bar{U} \subseteq O$. It remains to show (2).

$$\begin{aligned} \mathfrak{A}\bar{U}a \wedge U \subseteq P \subseteq O \wedge \mathfrak{K}P &\rightarrow \mathfrak{A}\bar{U}a \wedge \bar{U} \subseteq P \wedge \mathfrak{K}P \wedge P \subseteq O && (3)\\ &\rightarrow \mathfrak{A}Pa \wedge \mathfrak{K}P && (1),(4)\\ &\rightarrow Pa && (5) \end{aligned}$$

From that by (6): $\mathfrak{A}\bar{U}a \rightarrow \bar{U}a$.

From (0) - (4) one gets: (7) $\mathfrak{A}\bar{U} = \bar{U}$

<u>Proof:</u> (4), (1), (2) give $U \subseteq \mathfrak{A}\bar{U} \subseteq \bar{U} \subseteq O$. (1) applied to this yields: $\mathfrak{A}(\mathfrak{A}\bar{U}) \subseteq \mathfrak{A}\bar{U}$. According to (3) now $\bar{U} \subseteq \mathfrak{A}\bar{U}$.

Dual to (2) - (4), (7) one has the existence of a set $\bar{O}$ with:

(8) $\mathfrak{A}\bar{O} = \bar{O}$ (9) $Q \subseteq \mathfrak{A}Q \wedge U \subseteq Q \subseteq O \rightarrow Q \subseteq \bar{O}$ (10) $U \subseteq \bar{O} \subseteq O$

For $\mathfrak{A}P$ not depending on P this merely means comprehension. However for

$\mathfrak{A}P$ depending on P $\bar{U},\bar{O}$ are extremes for sets closed against $\mathfrak{A}$-iterations. This becomes transparent by the following transfinite approximation process for $\bar{U},\bar{O}$.

$\bar{U}$-approximation:

$$(11)\begin{cases} U_o =: U & \\ U_\alpha =: \mathfrak{A}U_\beta & \text{if } \alpha=\beta+1 \\ U_\lambda =: \bigcup_{\beta<\lambda} U_\beta & \text{if } \lambda \text{ lim-number}\end{cases} \qquad (12)\ \bar{U} =: \bigcup_\alpha U_\alpha$$

$\alpha,\beta,\ldots,\lambda,\ldots$ here denote ordinal numbers. - These U_α form an ascending sequence.

(13) $U \subseteq U_\alpha \subseteq U_{\alpha+1} \subseteq O$

Proof: Transfinite induction on α

1. $\alpha=0$: $U \subseteq U_o = U \subseteq \mathfrak{A}U = U_1 \subseteq O$ (11),(1)

2. $\alpha=\beta+1$: $U \subseteq U_\beta \subseteq U_{\beta+1} = U_\alpha \subseteq O$ according to induction hypothesis. With (1),(11) now: $U \subseteq U_\alpha = \mathfrak{A}U_\beta \subseteq \mathfrak{A}U_\alpha = U_{\alpha+1} \subseteq O$.

3. $\alpha=\lambda$ lim-number: By induction hypothesis:

$\gamma<\lambda \rightarrow U \subseteq U_\gamma \subseteq U_{\gamma+1} \subseteq O$. Thus

$\gamma<\lambda \rightarrow U \subseteq U_\gamma \subseteq U_{\gamma+1} \subseteq \bigcup_{\alpha<\lambda} U_\alpha = U_\lambda \subseteq O$ (11)

$\rightarrow U \subseteq U_\gamma \subseteq U_{\gamma+1} = \mathfrak{A}U_\gamma \subseteq \mathfrak{A}U_\lambda = U_{\lambda+1} \subseteq O$ (1),(11)

Consequently $U \subseteq \bigcup_{\gamma<\lambda} U_\gamma = U_\lambda \subseteq U_{\lambda+1} \subseteq O$. (11)

(14) $\alpha<\beta \rightarrow U \subseteq U_\alpha \subseteq U_\beta \subseteq O$

Proof: Transfinite induction on β

1. $\beta=0$: $\alpha<0 \rightarrow \curlywedge \rightarrow U \subseteq U_\alpha \subseteq U_\beta \subseteq O$

2. $\beta=\gamma+1$: $\alpha<\gamma+1 \rightarrow \alpha<\gamma \vee \alpha=\gamma$

$\rightarrow U \subseteq U_\alpha \subseteq U_\gamma \subseteq U_{\gamma+1} \subseteq O \vee U \subseteq U_\alpha \subseteq U_{\gamma+1} \subseteq O$ ind.hyp.,(13)

3. $\beta=\lambda$ lim-number: $\alpha<\lambda \rightarrow U \subseteq U_\alpha \subseteq \bigcup_{\alpha<\lambda} U_\alpha = U_\lambda \subseteq O$ (13),(11)

According to (14) the ascending U_α-sequence is contained in the power set of O and is therefore bounded in cardinality. Thus classically there exists an α with $U_\alpha = U_{\alpha+1}$.

(15) ξ =: least α with $U_\alpha = U_{\alpha+1}$

(16) $U_\alpha \subseteq U_\xi$ for arbitrary α

<u>Proof:</u> Transfinite induction on α

1. $\alpha=0$: $U_0 = U \subseteq U_\xi$ (11),(13)

2. $\alpha=\beta+1$: By induction hypothesis, (13): $U \subseteq U_\beta \subseteq U_\xi \subseteq O$. From this with (1),(11),(15): $U_\alpha = U_{\beta+1} = \mathfrak{A} U_\beta \subseteq \mathfrak{A} U_\xi = U_{\xi+1} = U_\xi$.

3. $\alpha=\lambda$ lim-number: According to induction hypothesis: $\beta<\lambda \longrightarrow U_\beta \subseteq U_\xi$. Thus $U_\alpha = U_\lambda = \bigcup_{\beta<\lambda} U_\beta \subseteq U_\xi$ (11).

(17) $\bar{U} = U_\xi$

<u>Proof:</u> (12),(16)

(2) - (4), (7) can now be proved as follows.

Ad (4): $U \subseteq \bar{U} \subseteq O$ by (17),(13)

Ad (2),(7): $\mathfrak{A}\bar{U} = \mathfrak{A} U_\xi$ (17),(13),(1)

$= U_{\xi+1} = U_\xi = \bar{U}$ (11),(15),(17)

Ad (3): On account of (12) it suffices to show $\mathfrak{A}Q \subseteq Q \wedge U \subseteq Q \subseteq O \longrightarrow U_\alpha \subseteq Q$ by transfinite induction on α.

1. $\alpha=0$: $\mathfrak{A}Q \subseteq Q \wedge U \subseteq Q \subseteq O \longrightarrow U_0 = U \subseteq Q$ (11)

2. $\alpha=\beta+1$: $\mathfrak{A}Q \subseteq Q \wedge U \subseteq Q \subseteq O \longrightarrow U \subseteq U_\beta \wedge U_\beta \subseteq Q \subseteq O$ ind.hyp.,(13)

$\longrightarrow U_\alpha = U_{\beta+1} = \mathfrak{A} U_\beta \subseteq \mathfrak{A} Q \subseteq Q$ (1),(11)

3. $\alpha=\lambda$ lim-number: $\mathfrak{A}Q \subseteq Q \wedge U \subseteq Q \subseteq O \longrightarrow \beta < \lambda \longrightarrow U_\beta \subseteq Q$ ind.hyp.

$\longrightarrow U_\lambda = \bigcup_{\beta<\lambda} U_\beta \subseteq Q$ (11)

Dual to this one has the descending <u>$\bar{O}$-approximation:</u>

$$(18)\ \begin{cases} O_0 =: O \\ O_\alpha =: \mathfrak{A} O_\beta & \text{if } \alpha=\beta+1 \\ O_\lambda =: \bigcap_{\beta<\lambda} O_\beta & \text{if } \lambda \text{ lim-number} \end{cases}$$

$$(19)\quad \bar{O} =: \bigcap_\alpha O_\alpha = O_\eta \quad \text{for sufficient large } \eta$$

These approximation processes can also be described set theoretically without ordinal numbers. Kuratowski [30] has used this to eliminate such applications of ordinal numbers. As an example we mention the corresponding version of Zermelo's second well-ordering proof [30], 87-88: the above approximation processes build up (exhaust) the set considered in forming a well-ordering along the way given by the axiom of choice.

The <u>induction spectrum</u> $\mathfrak{J}$ of $\mathfrak{A}$ between U,O be $\{P \mid U \subseteq P \subseteq O \wedge \mathfrak{A}P=P\}$. By Tarski [62], Theorem 1 (starting from the complete $\cap,\cup,\subseteq$-lattice $\mathfrak{P}$ of all sets between U,O) $\mathfrak{J}$ is a complete $\subseteq$-lattice. Herein $\bar{U}$ is the 0- and $\bar{O}$ the 1-element. However in general $\mathfrak{J}$ is no sublattice of $\mathfrak{P}$, so that inf, sup in $\mathfrak{J}$ in general don't coincide with $\bigcap, \bigcup$. But because for in $\mathfrak{J}$ comparable elements $\subseteq$ holds inf, sup can be approximated in $\mathfrak{J}$ as follows.

Consider $\mathfrak{M} \subseteq \mathfrak{J}$. Since $\mathfrak{J}$ is a complete $\subseteq$-lattice, so for $M \in \mathfrak{M}$ holds:
(20) $U \subseteq \inf \mathfrak{M} \subseteq \bigcap_{\mathfrak{M}} \subseteq M \subseteq O, \quad U \subseteq M \subseteq \bigcup_{\mathfrak{M}} \subseteq \sup \mathfrak{M} \subseteq O$. With (1) therefore

$$U \subseteq \mathfrak{A}(\bigcap_{\mathfrak{M}}) \subseteq \mathfrak{A}M = M \;,\; M = \mathfrak{A}M \subseteq \mathfrak{A}(\bigcup_{\mathfrak{M}}) \subseteq O$$

$$(21) \quad U \subseteq \mathfrak{A}(\bigcap_{\mathfrak{M}}) \subseteq \bigcap_{\mathfrak{M}} \subseteq O \;,\; U \subseteq \bigcup_{\mathfrak{M}} \subseteq \mathfrak{A}(\bigcup_{\mathfrak{M}}) \subseteq O$$

and finally

$$U \subseteq P_1 \subseteq P_2 \subseteq \bigcap_{\mathfrak{M}} \subseteq O \rightarrow U \subseteq \mathfrak{A}P_1 \subseteq \mathfrak{A}P_2 \subseteq \mathfrak{A}(\bigcap_{\mathfrak{M}}) \subseteq \bigcap_{\mathfrak{M}} \subseteq O \qquad (21)$$

$$U \subseteq \bigcup_{\mathfrak{M}} \subseteq P_1 \subseteq P_2 \subseteq O \rightarrow U \subseteq \bigcup_{\mathfrak{M}} \subseteq \mathfrak{A}(\bigcup_{\mathfrak{M}}) \subseteq \mathfrak{A}P_1 \subseteq \mathfrak{A}P_2 \subseteq O \qquad (21)$$

Thus for the domains U, $O_1 =: \bigcap_{\mathfrak{M}}$ and $U_1 =: \bigcup_{\mathfrak{M}}$, O a relation analogous to (1) holds.
Consequently we can apply the above transfinite approximation processes. The so obtained $\bar{O}_1$, $\bar{U}_1$ are the greatest resp. least element of $\mathfrak{J}$ in the domain U, O_1 resp. U_1, O. Therefore by (20): $\sup \mathfrak{M} = \bar{U}_1$, $\inf \mathfrak{M} = \bar{O}_1$.

We now change over to the intuitionistic point of view. Here the circumstances are different than before because arbitrary comprehensions are up to now not accepted offhand (compare Troelstra [63], §4,15) (To continuity contradicts e.g. the 00-comprehension formulated in the functional language with $\underline{\Omega}$: Howard-Kreisel [19], 330 with (3.15)), and effective well-orderings are only available very restrictedly (only segments of the second number class; Troelstra [63], § 14). Yet here one can (and must) as well speak of generalized inductive definitions; only the general comprehension proof from above should be replaced from case to case by better arguments. - In the following the most important intuitionistic and also here relevant cases are treated, namely trees, ordinal numbers and continuous functionals. We shall begin with Brouwer's bar induction which is very closely related to these ideas. We are not so much interested in the formal-axiomatic aspect as rather in the intuitive motivation from where the rest can be surveyed and at which generalizations in the next chapter shall be resumed.

Brouwer's bar induction BI^{o}_{D} over countable trees

Notation: α for choice sequences over numerals

(1) Hyp.: $P(u^{oo},v^{o})$ sequence-extensional in u

(2) Hyp.: $P(<\alpha 0,\ldots,\alpha(x-1)>,x) \vee \neg P(<\alpha 0,..,\alpha(x-1)>,x)$, i.e. P is decidable in the spread over the numerals.

(3) Hyp.: $\bigwedge \alpha^{oo} \bigvee x^{o}\ P(<\alpha 0,\ldots,\alpha(x-1)>,x)$

From (2),(3): (4) $\bigwedge \alpha \bigvee x^{o} \{ \underbrace{P(<\alpha 0,..,\alpha(x-1)>,x) \wedge \bigwedge y^{o}<x \neg P(<\alpha 0,..,\alpha(y-1)>,y)}_{P^{*}(<\alpha 0,..,\alpha(x-1)>,x)\ \equiv :}$

In the full infinitely branched spread over the numerals (1),(4) fix a decidable tree $\mathcal{L}^{*}$, i.e. to each path in it there is exactly one (decidable) P^{*}-bar.
An intuitionistic direct proof of (4) has the form of a well-ordered countable tree $\mathcal{L}$ of argumentations with finite or infinite ramifications. $\mathcal{L}$ is well-describable.
$\mathcal{L}$, $\mathcal{L}^{*}$ carry the same bar informations for P. Now the thesis is layed down that the direct argumentation $\mathcal{L}$ ultimately amounts to the fact that - seen in the tree $\mathcal{L}^{*}$ - the property P can be proved for all strings with initial segments in $\mathcal{L}^{*}$ (and thus actually for each string of the spread) in starting with the outermost direct decidable P^{*}-bar points and then proceeding inwards by inferences using (1). Thus this thesis says that the (direct) argumentation $\mathcal{L}$ can be standardized in the form $\mathcal{L}^{*}$ given by the result; in other words $\mathcal{L}^{*}$ is like $\mathcal{L}$ a well describable proof tree in which starting with the "axioms" one can effectively proceed by the infinite inference relations to the "end formula" (induction over $\mathcal{L}^{*}$).

To the proof tree $\mathcal{L}^{*}$ we now refer the remaining premises of BI^{o}_{D}.

(5) Hyp.: $P(<\alpha 0,\ldots,\alpha(x-1)>,x) \longrightarrow Q(<\alpha 0,\ldots,\alpha(x-1)>,x)$

(6) Hyp.: $\bigwedge z^{o}\ Q(<\alpha 0,\ldots,\alpha(x-1),z>,x') \longrightarrow Q(<\alpha 0,\ldots,\alpha(x-1)>,x)$

Because of (5) Q holds at the "axioms" of $\mathcal{L}^{*}$. By (6) Q is hereditary at the infinite ramifications ("inferences") of $\mathcal{L}^{*}$. Thus by induction over $\mathcal{L}^{*}$: $Q(<>,0)$.

Remark:

1. In short: BI^{o}_{D} means induction over the proof tree given by (4).

2. BI^{o}_{D} is impredicative because for Q there may stand propositions which

use $\mathfrak{B}^*$-properties so that $\mathfrak{B}^*$-induction over $\mathfrak{B}^*$-properties may occur.

Representatives for ordinals of the second number class and continuous functionals of type 0(00) are only different dictions over countable trees.

Inductive generation of the class $\mathfrak{B}$ of the countable trees

$\mathfrak{B}$ is intuitionistically defined as the class of trees which are built up directly according to the following principles:

(a) $B \equiv \bullet \longrightarrow B \varepsilon \mathfrak{B}$ (B consists of one node $\bullet$)

(b) $B \equiv$ 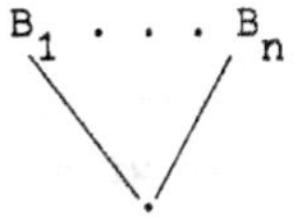

$\wedge \bigwedge_{i=1}^{n} B_i \varepsilon \mathfrak{B} \longrightarrow B \varepsilon \mathfrak{B}$

(c) $B \equiv$ 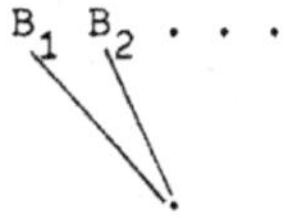

$\wedge \bigwedge i \; B_i \varepsilon \mathfrak{B} \longrightarrow B \varepsilon \mathfrak{B}$

At this the specification of the sequences B_i can still be subjected to certain conditions. So Heyting [14], 188 and Troelstra [63], 76 allow only lawlike sequences. Without reference to Church's thesis one can further distinguish:

1) only recursive sequences B_i: The so defined recursive trees can be characterized by a number.

2) only computable sequences B_i: Computable trees

However Brouwer admitted in [1], 451 and [3], 10 greatest possible liberty and emphasized this explicitly in [2]: "jedenfalls habe ich ... seit geraumer Zeit betont, daß eine beliebige stetige Funktion genauso "im freien Werden" entsteht wie ein beliebiger Punkt des Kontinuums." It is exactly this possibility we will use in the following.

3) choice sequences B_i: Choice trees; contrary to the fixed trees under 1), 2) they are always freely growing over the so far generated subclass of trees.

In the following "tree" stands for each of these kinds of trees respectively.

Each $B \varepsilon \mathcal{L}$ represents his proof as tree of $\mathcal{L}$ directly. Further with the abbreviation $\mathfrak{A}(\mathcal{L},B)$ for the disjunction of the premises in (a) - (c) we have

(i) $\mathcal{I} \subseteq \mathcal{J} \longrightarrow \mathfrak{A}(\mathcal{I},B) \longrightarrow \mathfrak{A}(\mathcal{J},B)$

(ii) $\mathfrak{A}(\mathcal{L},B) \longrightarrow B \varepsilon \mathcal{L}$

and the following "(generation) induction over $\mathcal{L}$"

(iii) $\bigwedge C\{\mathfrak{A}(\mathcal{I},C) \longrightarrow C \varepsilon \mathcal{I}\} \longrightarrow \mathcal{L} \subseteq \mathcal{I}$

because in a tree $B \varepsilon \mathcal{L}$ by $\mathfrak{A}(\mathcal{I},C) \longrightarrow C \varepsilon \mathcal{I}$ the construction of B according to (a) - (c) can be followed up by steps in proving that every subtree of B and finally also B is in $\mathcal{I}$.

(i) - (iii) say that $\mathcal{L}$ was generated by an intuitionistically justified generalized inductive definition. $\mathcal{L}$ is the least class with (ii). This minimum condition (iii) intuitionistically determines by (ii) the direct $\mathcal{L}$-generation according to (a) - (c) so that (ii), (iii) also conversely define unambiguously the generation process given in the beginning.

According to generation $\mathcal{L}$ contains only countable trees. Also conversely all countable trees are in $\mathcal{L}$. This can be drawn from the previously given argumentation for BI_D^o. After that each (provable) countable tree is a proof tree. Induction on it yields by (a) - (c) its membership to $\mathcal{L}$. - Thus $\mathcal{L}$ justly bears the above given name "class of the countable trees": $\mathcal{L}$ = class of the proof trees. Over each $B \varepsilon \mathcal{L}$ BI_D^o is valid.

Ordinal numbers

$\mathcal{L}$ together with the decidable subspecies of $B \varepsilon \mathcal{L}$ represent intuitionistically the class of the well-ordered species; every tree $B \varepsilon \mathcal{L}$ is well-ordered by the decidable Brouwer-Kleene ordering at which a node k_1 is "less" than a node k_2 iff the path up to k_1 is left of the path up to k_2 or the latter is contained in the former. Order-isomorphic abstraction gives the ordinals of the second number class. Corresponding to the sort of the trees we distinguish between recursive, computable and choice ordinals. For further details see Troelstra [63],§14.

Continuous functionals of type 0(00)

Starting point are the countable trees $B \in \mathcal{L}^{\infty}$ which contain only infinite ramifications. $\mathcal{L}^{\infty}$ is generated analogously to $\mathcal{L}$ when (b) is left out.

In such a $B \in \mathcal{L}^{\infty}$ the nodes at the branchings be numbered consecutively from the left to the right and to each maximal path in B (in other words to each end point in B) a number be associated. Correspondences of this kind determine exactly the extensional and effective continuous functionals F of type 0(00) by associating to each choice sequence α over numerals the value which was previously given to their maximal initial segment in B. We designate this functional class by $\tilde{\mathcal{R}}_1$ (continuous functionals over $\tilde{\mathcal{R}}_0$ =: {numerals}) and distinguish again between recursive, computable and continuous choice functionals according to whether the specification of the tree $B \in \mathcal{L}^{\infty}$ and the number correlation over B uses only recursive, computable or also choice operations.

In the functional language still it is possible to comprise the tree construction and the number correlation to one inductive process by producing directly a tree at nodes of which there are continuous functionals:

$$\gamma(\tilde{\mathcal{R}}, F^{0(00)}) \quad \equiv: \bigvee \text{numeral } z \; F = \lambda\alpha^{00} z \vee \bigwedge \text{numerals } y \; \lambda\alpha^{00} \; F(\langle y, \alpha 0, \alpha 1, \ldots \rangle) \in \tilde{\mathcal{R}}$$

whereby for the dependence of $\lambda\alpha F(\langle y, \alpha 0, \alpha 1, \ldots\rangle)$ in y^0 again recursiveness, computability, free choice or the like may be required.

(i) $\tilde{\mathcal{R}} \subseteq \mathcal{L} \rightarrow \gamma(\tilde{\mathcal{R}}, F) \rightarrow \gamma(\mathcal{L}, F)$

Analogous as previously for $\mathcal{L}$ we have

(ii) $\gamma(\tilde{\mathcal{R}}_1, F) \rightarrow F \in \tilde{\mathcal{R}}_1$

(iii) $\bigwedge F\{\gamma(\mathcal{L}, F) \rightarrow F \in \mathcal{L}\} \longrightarrow \tilde{\mathcal{R}}_1 \subseteq \mathcal{L}$

and conversely (ii), (iii) determine the above intuitively introduced $\tilde{\mathcal{R}}_1$ unambiguously on account of the directness of the generation process following from the minimality.

For $F \in \tilde{\mathcal{R}}_1$ according to generation induction $\bigwedge\alpha \bigvee ! x^0$ {in the F-specification exactly for all choice sequences having the initial segment $\langle \alpha 0, \ldots, \alpha(x-1)\rangle$ the same value is unambiguously given} holds. And conversely such F with $\bigwedge\alpha \bigvee ! x^0 \{\ldots\}$ are in $\tilde{\mathcal{R}}_1$ because according to the above BI_D^0-analysis from $\bigwedge\alpha \bigvee ! x^0 \{\ldots\}$ the existence of an infinitely ramified proof tree with exactly the same bar properties follows and an induction over this tree using the generation instructions for continuous functionals finally yields $F \in \mathcal{R}_1$.

Analogous as for numerals one can deal with every effective denumerable set M and define continuous functionals over M-choice sequences. Also finitely many arguments can be treated analogously.

XII. Generalization of bar induction BI_D and the inductive generation processes to trees over species

In the following we make the attempt to motivate (the classically valid) bar induction BI_D and the inductive generation processes for trees over arbitrary species M as perhaps even intuitionistically acceptable generalizations of Brouwer's bar induction BI_D^o and the inductive generation processes above. Such generalized BI_D we need in a natural way for the computability proof of higher bar recursions.

So let a non-empty species M be given, i.e. a totality of previously generated elements for which a certain definition property is provable. It is not possible to copy the classical notion of "higher spread or tree over M" in full intuitionistically because we connect with this notion a well-ordering of the ramifications and effective well-orderings are not available for arbitrary M. However intuitionistically one can operate with such generalizations of the previous concepts in a necessary and natural restriction compared to the classical version. This will be performed now in more detail; again we begin with BI_D.

General BI_D for trees over species M

Notation: "α over M" for freely chosen sequences α of M-elements for which the M-membership has been proven before each choice act.

(1) Hyp.: $P(u^{\mu o},v^o)$ sequence-extensional in u $\quad$ (μ =: type of M)

(2) Hyp.: $P(<\alpha 0,\ldots,\alpha(x-1)>,x) \vee \neg P(<\alpha 0,\ldots,\alpha(x-1)>,x)$ for α over M,

i.e. P is decidable in the "M-spread".

(3) Hyp.: $\bigwedge \alpha$ over M $\bigvee x^o\ P(<\alpha 0,\ldots,\alpha(x-1)>,x)$

From (2),(3):

(4) $\bigwedge\alpha$ over M $\bigvee x^o \underbrace{\{P(<\alpha 0,..,\alpha(x-1)>,x) \wedge \bigwedge y^o<x \neg P(<\alpha 0,..,\alpha(y-1)>,y)\}}_{P^*(<\alpha 0,..,\alpha(x-1)>,x)\ \equiv:}$

An intuitionistic direct proof of (4) has the form of a well-describable and well-ordered countable tree $\mathcal{L}$ of argumentations with finite or infinite ramifications. Because in general the extension of M cannot be surveyed effectively, so generally the constructive unique correlation of numbers to choice sequences α over M and their M-membership proofs given in (4) can only be based on certain countable classifying uniformizations. This can be taken effectively from the demonstration $\mathfrak{L}$ and can be standardized as follows in a countable "correlation tree" $\mathfrak{L}^*$ with finite or infinite branchings at the nodes of which there are species of M-elements together with their M-proofs according to the $\mathfrak{L}$-argumentation so that a ramification in $\mathfrak{L}^*$ covers all M-elements: a choice sequence α over M is P^*-valued according to (4) by its $\mathfrak{L}$-minimal initial segment in $\mathfrak{L}^*$ so that all 'neighborhoods' to it given by $\mathfrak{L}^*$ have in a $\mathfrak{L}$-uniform manner the same P^*-valuation. For effective denumerable M these uniformities can be dissolved to an effective denumerable spread $\mathfrak{L}^*$ at the nodes of which there is always only one M-element. This leads to the previously treated Brouwer-BI_D^o.

Thus $\mathfrak{L}, \mathfrak{L}^*$ carry the same bar informations for P. - Now the thesis is layed down that the direct argumentation $\mathfrak{L}$ ultimately amounts to the fact that - seen in the tree $\mathfrak{L}^*$ - the P-property can be proved for all choice sequences α over M with initial segments in $\mathfrak{L}^*$ (and thus actually for all) in starting with the outermost, in each individual case direct decidable P^*-bar points and then proceeding inwards by inferences using (1) and the $\mathfrak{L}$-uniformities. Thus this thesis says that $\mathfrak{L}$ (on account of the uniformities) can be standardized in $\mathfrak{L}^*$-form; so $\mathfrak{L}^*$ is also a well-describable countable proof tree in which starting with the "axioms" one can effectively proceed by the finite or denumerable infinite "inference relations" to the "end formula" (induction over $\mathfrak{L}^*$).

To this proof tree $\mathfrak{L}^*$ we now refer the remaining premises of general BI_D.

(5) Hyp.: $P(<\alpha 0,..,\alpha(x-1)>,x) \longrightarrow Q(<\alpha 0,..,\alpha(x-1)>,x)$

(6) Hyp.: $\bigwedge z\varepsilon M\ Q(<\alpha 0,..,\alpha(x-1),z>,x') \longrightarrow Q(<\alpha 0,..,\alpha(x-1)>,x)$

for α over M

Because of (5) Q holds at all 'points' of the $\mathfrak{L}^*$-axioms. Since the $\mathfrak{L}^*$-branchings cover full M by (6) Q is hereditary at the $\mathfrak{L}^*$-inferences. Thus by induction over $\mathfrak{L}^*$: $Q(<>,0)$.

Remark:

1. General BI_D is impredicative for the same reasons as previously BI_D^o.

2. Thus by uniformizations general BI_D becomes a Brouwer-BI_D^o on a higher abstraction level. The classical higher spread over M is intuitionistically on account of the uniformizations arising with the con-

structive treatment of higher connections pushed together to a countable spread over M-subspecies.

Inductive generation of higher 'trees' and 'ordinal numbers'

To generate intuitionistically higher trees - perhaps starting with the class $\mathfrak{B}$ of the countable trees - leads to the following. Again one begins with single $B \in \mathfrak{B}$ and then proceeds by effective correlations of previously generated higher trees to the elements of a subspecies γ of $\mathfrak{B}$.

As in the above analysis of general BI_D such an effective correlation *may* lead by uniformizations to a denumerable row of γ-subspecies to the elements of which in a uniform manner the same higher tree is associated respectively. So altogether one may get - e.g. in cases as above - to a *countable correlation tree* in γ-subspecies from which the higher tree by the so fixed uniformity results. The ramifications herein can still be refined by ordinals of the second number class; but this leads to nothing new in principle. - Similar facts hold for higher correlation trees.

Analogous to this are the properties of the corresponding generalized 'ordinal numbers'. It is true that in countable correlation trees again there is effectively the (possibly ordinal generalized) Brouwer-Kleene well-ordering. But even for such higher trees this constitutes *only a partial ordering* (no distinction between uniform treated nodes).

The general situation does not change for arbitrary species M instead of $\mathfrak{B}$. Over each higher tree general BI_D is valid. - Obviously the notions "tree" and "ordinal number" can intuitionistically be generalized only in this way.

Continuous number-functionals on trees over species M

Here trees B_M over M with countable correlation tree the branchings of which cover all M-elements are taken as a basis. Analogous to chapter XI here one uniformly has the number correlation over the (countable) correlation tree of B_M. Several arguments are treated correspondingly.

In generalization of the already introduced $\mathfrak{K}_0$, $\mathfrak{K}_1$ we especially mention the classes $\mathfrak{K}_{i+1}$ of continuous number-functionals on the trees over $\mathcal{R}_i$. Similarly one has higher-valued functionals on corresponding higher correlation trees.

Now, in what way these means can be intuitionistically employed for functional interpretation, i.e. for the computability proof of a suitable extension of T∪BR? Let us consider the so far made attempts in this direction.

If one starts with the countable resp. continuous functionals in Kleene [20], Kreisel [23] then one has to prove in this model the existence of the bar recursive functionals. This leads to the difficulties described in Kreisel [27], §4.

Tait announced 1966 in [54], 201 a treatment of higher bar recursions and restricted this 1967 in [55], 197 to the types $\Theta_0 \equiv :0$, $\Theta_{m+1} \equiv :0(\Theta_m 0)$. These types point at an application of the above classes $\mathcal{R}_m$ of continuous functionals. An extension of the so far considered functional domains by these continuous functionals of $\mathcal{R}_m$ brings along the following difficulties for the computability proof. Because the new functional domain must be closed against applications the domain of definition of the $\mathcal{R}_m$-functionals has to be expanded to this. However this must be done so that extensionality and continuity which are needed for the computability proof are preserved. This point ist not self-evident, for the general definition of the $\mathcal{R}_m$-functionals actually does not care about any connections in the argument domain. - Whereas this can be done for type 0 and even for the types 0...0 in making the usual interpretation precise (but will be more complicated than the above given treatment) already for type 0(00) one has to proceed in another way. Possibly the same principle on which the countable functionals are based will suffice here. By extending this to several arguments and using the Howard-Kreisel method from above, one would get a BR-foundation for the types: 0; $\alpha \Rightarrow \alpha 0$; $\alpha_1, \ldots, \alpha_n \Rightarrow 0(\alpha_1 0)\ldots(\alpha_n 0)$; but already 0(0(00)) is not covered by this. A demonstration of these announced results has - to the author's knowledge - not been published by Tait till now.

From this we draw the conclusion that the needed closure of a model containing T∪BR against choice sequences is <u>not</u> achieved in the simplest way by building up models which at least in some respects try to offer a complete intuitionistic functional theory because by this one gets to domains and problems which seem to go far beyond the concrete meta-mathematical problem under investigation. A in this sense meta-mathematical orientated and naturally constructed model for full $T \cup BR$ gives the following chapter.

XIII. A model for T∪BR

It was noticed in chapter X that the argumentation carries over to higher bar recursions provided the underlying functional domain is closed against choice sequences of higher type. Let us try to achieve this from the conceptual side in simply closing off the T∪BR-operations and the formation of choice sequences in themselves by a tree-like term construction.

$\mathfrak{M}$ be the functional term class which is directly generated by the following instructions:

(a) $\varphi \equiv 0 \longrightarrow \varphi^{0} \varepsilon \mathfrak{M}$

(b) $\varphi \equiv ' \longrightarrow \varphi^{00} \varepsilon \mathfrak{M}$

(c) φ according to $(BR)^{\alpha} \longrightarrow \varphi \varepsilon \mathfrak{M}$

(d) φ according to (T1) $\longrightarrow \varphi \varepsilon \mathfrak{M}$

(e) $\psi \varepsilon \mathfrak{M} \wedge \varphi$ from ψ according to (T2) $\longrightarrow \varphi \varepsilon \mathfrak{M}$

(f) $\psi, \chi \varepsilon \mathfrak{M} \wedge \varphi$ from ψ, χ according to (T3) $\longrightarrow \varphi \varepsilon \mathfrak{M}$

(g) $\psi, \chi \varepsilon \mathfrak{M} \wedge \varphi$ from ψ, χ according to (T4) $\longrightarrow \varphi \varepsilon \mathfrak{M}$

(h) $\varphi_0^{\alpha\alpha_1}, \varphi_1^{\alpha_1} \varepsilon \mathfrak{M} \wedge \varphi \equiv (\varphi_0 \varphi_1)^{\alpha} \longrightarrow \varphi \varepsilon \mathfrak{M}$

(i) $\varphi \equiv$ choice sequence $<\varphi_0^{\alpha}, \varphi_1^{\alpha}, \ldots> \wedge \bigwedge$ nat i $\varphi_i \varepsilon \mathfrak{M} \longrightarrow \varphi^{\alpha 0} \varepsilon \mathfrak{M}$

With the abbreviation $\mathfrak{A}(\mathfrak{M}, \varphi)$ for the disjunction of the in $\mathfrak{M}$-magnitudes $\bigvee$-closed premises of (a) - (i) intuitionistically - just like before - the generalized inductive definition holds:

$$\mathfrak{M}_1 \subseteq \mathfrak{M}_2 \rightarrow \mathfrak{A}(\mathfrak{M}_1, \varphi) \longrightarrow \mathfrak{A}(\mathfrak{M}_2, \varphi)$$

$$\mathfrak{A}(\mathfrak{M}, \varphi) \rightarrow \varphi \varepsilon \mathfrak{M}$$

$$\bigwedge \psi \{ \mathfrak{A}(\mathfrak{N}, \psi) \rightarrow \psi \varepsilon \mathfrak{N} \} \longrightarrow \mathfrak{M} \subseteq \mathfrak{N}$$

This naturally constructed functional domain $\mathfrak{M}$ obviously forms the minimal model which is closed against the T∪BR-operations and choice sequences. Behind the $\mathfrak{M}$-generation there is exactly the formation of choice trees. Therefore each $t \varepsilon \mathfrak{M}$ reflects the proof of its membership to $\mathfrak{M}$ directly.

$\mathfrak{M}$ is - up to μ which however can be adjoined unconstrained as in (10.12) - the most extensive functional domain of the kind considered in chapter X. Thus the theorems proved there now hold for $\mathfrak{M}$. Especially from this one gets the extensional computability of every term $t \varepsilon \mathfrak{M}$ by a generation induction over its term construction tree:

(a),(b): (10.2)

(c): The extensional computability of the bar recursive functionals $\varphi \equiv \phi^{(\alpha)}$ follows quite analogously as in the proof of (10.14) on page 112-115 when the following modifications are made.
τ_i, τ_{i+1} are replaced by α, $\alpha 0$. Now naturally a general bar induction $BI_D^{(\alpha)}$ is used which runs over the "higher trees" given by the extensionally computable $\mathfrak{M}$-functionals of type $0(\alpha 0)$ according to (10.11) in the "higher spread" over the (predicative) species of extensionally computable type α-terms of $\mathfrak{M}$. Therefore instead of "ics $f^{\tau_{i+1}}$", "ics g^{τ_i}" now one has to take "cs $f^{\alpha 0}$ over the species of the extensionally computable type α-terms of $\mathfrak{M}$" and "g^{α} out of the species of the extensionally computable type α-terms of $\mathfrak{M}$". - The application of (10.13) on page 115 falls away.

(d) - (g): like (10.7)

(h): with (10.1)

(i): as in the proof of (10.5), $(*_3)$

Thus constructively - in the sense of the means used - it is proved:

(13.1) $\mathfrak{M}$ is extensionally computable relative to itself.

Hence it follows with (10.15):

(13.2) T∪BR is extensionally computable relative to itself.

Together with (8.2), (10.9) this yields the final result:

(13.3) The functional interpretation of classical $(AC)^0$-, (ωAC)-analysis is constructively given in T∪BR. Corresponding statements hold for the functional interpretation in the narrower sense of the previously treated intuitionistic approximated theories.

Remark:
If the formation of choice sequences in (i) is restricted to certain types then correspondingly one obtains submodels for T plus bar recursive functionals of these types.

The means used in the given proofs are certainly intuitionistic with the only exception of the to species generalized bar induction BI_D. In the author's opinion the intuitive motivation given to general bar induction BI_D over species in chapter XII as an on a higher abstraction level in uniformization rooted Brouwer bar induction BI_D^o is also intuitionistically acceptable. The above explanations then would yield an intuitionistic functional interpretation - and thus (among other things, see (10.9)) also an intuitionistic consistency proof - of classical $(AC)^o$-, (ωAC)-analysis.

Concluding let us mention that the whole argumentation with the corresponding change in the interpretation - and that's just the point here - is also classically valid.

XIV. On the bar recursive model of classical analysis and the general bar induction over species

In what natural and as minimal as possible deductive framework can the above consistency proof of classical analysis (i.e. ZFC^- Zermelo-Fraenkel set theory without power set axiom) be formalized? According to our proof-theoretic experience over a suitable language Heyting's arithmetic plus certain induction principles may be best.

It is not possible to carry out here the exact formalization process because of its considerable technical display. But we will describe the most important steps; the reader then may verify the details he is interested in.

First in a finitary way classical analysis was interpreted in the functional calculus $T \cup BR$. This can be described in the usual Heyting arithmetic including the primitive recursive functions.

The proof of the extensional computability exceeded finitary methods. Bar recursion made it necessary to extend $T \cup BR$ to the intuitionistic model $\mathfrak{M}$ which is closed against choice sequences in all types and is therefore not denumerable. The $\mathfrak{M}$-elements are well-founded countable trees; below it is described how these trees can be represented by functionals of type 00. Thus a language $\mathfrak{T}$ is needed which deals with numbers, number-theoretic functions, choice sequences over numbers, primitive recursive functional relations between these, $\overset{0}{=}$ and λ-abstraction (Kleene-Vesley [22]); these means for which extensionality can be proved will also be sufficient.

At first we are interested in the local consistency, i.e. the demonstration that single proof figures do not contain any contradiction. This reduces to the computability proof of the single functionals used in the functional interpretation. For this it is necessary that $\mathfrak{M}$ is closed against the $T \cup BR$- and choice-sequence operations; this is the content of the first condition of the generalized inductive definition for $\mathfrak{M}$. It is further required that the $\mathfrak{M}$-elements are built up tree-like using only these operations; this is the content of the second condition of the generalized inductive definition for $\mathfrak{M}$ (recursion over the construction of the $\mathfrak{M}$ -elements). Below it is shown that the $\mathfrak{M}$-membership can be described by a formula of $\mathfrak{T}$ and that the first

$\mathfrak{M}$-condition is provable in Heyting arithmetic plus T and the second $\mathfrak{M}$-condition follows therein with additional Brouwer bar induction BI_D^0.

The reduction instructions of the computation are coded as primitive recursive formation instructions for $\mathfrak{M}$-functions from preceding $\mathfrak{M}$-functions. Every (finite) computation can again be represented as a function. - Extensional computability and computational equality can be expressed in $\mathcal{T}$ only for single types because of their quantifier structure.

The extensional computability relative to $\mathfrak{M}$ of the single T-, choice sequence and μ-operations and the proof that from the computability of the finitely many $\mathfrak{M}$-elements occurring in a demonstration the local consistency follows (to this in (10.8) the valuation is relativized to the non-empty computable part of $\mathcal{F}(\mathfrak{M})$; then it is only used that the finitely many functionals of the demonstration (their predecessors included) are computable relative to $\mathcal{F}(\mathfrak{M})$.) thus can be given over $\mathcal{T}$ in Heyting arithmetic plus T, BI_D^0. Only the bar recursive functionals need the higher special bar induction $(hsBI)_D^{00}$ over the non-empty species S of the extensional computable $\mathfrak{M}$-terms of the type considered. This can be formulated in $\mathcal{T}$ as follows.

2-place functions: $\alpha^{00}(x^0,y^0) \equiv: \alpha(\langle\langle x,y\rangle\rangle)$

for a primitive recursive pairing function $\langle\langle,\rangle\rangle$

Finite sequence of functions:

$$\overline{\alpha,x}(y^0,z^0) \equiv: \begin{cases} \alpha^{00}(y,z) & \text{if } y<x^0 \\ 0 & \text{otherwise} \end{cases}$$

Chaining of a link to a finite sequence of functions:

$$\overline{\alpha,x}*\beta^{00}(y^0,z^0) \equiv: \begin{cases} \alpha^{00}(y,z) & \text{if } y<x^0 \\ \beta z & \text{if } y=x \\ 0 & \text{otherwise} \end{cases}$$

(H0) $\bigvee \beta^{00} S(\beta)$

(H1) $\bigwedge \alpha^{00}\{\bigwedge x^0 S(\lambda z^0 \alpha(x,z)) \longrightarrow \bigvee y^0 A(\overline{\alpha,y};y)\}$

(H2) $\bigwedge \alpha^{00},x^0\{\bigwedge y^0<x S(\lambda z^0 \alpha(y,z)) \longrightarrow A(\overline{\alpha,x};x) \vee \neg A(\overline{\alpha,x};x)\}$

(H3) $\bigwedge \alpha^{00},x^0\{\bigwedge y^0<x S(\lambda z^0 \alpha(y,z)) \wedge A(\overline{\alpha,x};x) \longrightarrow B(\overline{\alpha,x};x)\}$

(H4) $\bigwedge \alpha^{00},x^0\{\bigwedge y^0<x S(\lambda z^0 \alpha(y,z)) \wedge \bigwedge \beta^{00}(S(\beta) \rightarrow B(\overline{\alpha,x}*\beta;x')) \longrightarrow B(\overline{\alpha,x};x)\}$

$(\mathrm{hsBI})_D^{oo}$: $(H0) \wedge \ldots \wedge (H4) \longrightarrow B(\sigma;0)$

$(\mathrm{hsBI})_D^{oo}$-rule: $\dfrac{(H1),\ldots,(H4)}{(H0) \longrightarrow B(\sigma;0)}$

Below it will be shown that the $(\mathrm{hsBI})_D^{oo}$-rule is deductive equivalent to $(\mathrm{hsBI})_D^{oo}$ (what similarly also holds for the Markov-principle) and that BI_D^o is derivable from $(\mathrm{hsBI})_D^{oo}$. This altogether yields the following result:

(I) The local consistency of classical (ωAC)-analysis resp. T$\cup$BR is provable in the language $\mathcal{L}$ by Heyting arithmetic plus T, $(\mathrm{hsBI})_D^{oo}$-rule; that is, for every single proof of classical (ωAC)-analysis resp. T$\cup$BR it can be shown in Heyting arithmetic plus T, $(\mathrm{hsBI})_D^{oo}$-rule that it does not lead to 0=1. Because on the other hand also a λ-proof of classical (ωAC)-analysis or T$\cup$BR could be directly expressed in this way within this subsystem (see remark to (VII)) so Heyting arithmetic plus T, $(\mathrm{hsBI})_D^{oo}$-rule has exactly the same proof-theoretic strength as classical (ωAC)-analysis or T$\cup$BR.

Thus T$\cup$BR gives exactly the (total) functions of type 0...0 which can be proved general recursive in classical analysis; T$\cup$BR also contains all $\bigwedge\bigvee$-correlations of classical analysis and further in this domain continuous functional operations.

The full (global) consistency of classical (ωAC)-analysis follows from this by adding the constructive instance of the ω-rule for the proof predicate of (ωAC)-analysis previously formulated in (I). The constructivity is established in the fact that the proof of the local consistency was given uniformly for each individual case by an effective homogeneous method.

(II) The consistency of classical (ωAC)-analysis is provable in the language $\mathcal{L}$ by Heyting-arithmetic plus T, $(\mathrm{hsBI})_D^{oo}$-rule and a constructive application of the ω-rule.

The difference between (II) and (I) corresponds exactly to the situation in Gödel's underivability theorems.

We still have to treat in more detail the generalized inductive definition of $\mathfrak{M}$ and $(\mathrm{hsBI})_D^{oo}$; next to $\mathfrak{M}$.

The $\mathfrak{M}$-elements are countable choice trees at the end nodes of which in preserving the type relations there are the basic functionals 0, ', $(BR)^{\alpha}$, (T1); to the ramifications corresponds unary a (T2)-formation, binary an application of two functionals or a (T3)- or (T4)-formation and infinitary a formation of a choice sequence.

The coding by prime factorization is based on the following one-to-one denumerations:

a) Denumeration of the finite types by natural numbers ≥ 0

b) Denumeration of the basic functionals by natural numbers ≥ 1

c) Denumeration of all ordered n-tupels $<z_1,\dots,z_n>$ (z_i numerals) with $<> =: 0$

For simplicity in the following <...> is sometimes identified with its denumeration number. Also the usual sequence operations (linkage etc.) are used.

Now a countable choice tree is represented by a choice function f of type 00 by specifying whether an initial segment of a path $<z_1,\dots,z_n>$ is inside or outside the tree. - Here every tree node f(<...>) (<> = origin of the tree) moreover carries the following informations:

a)

$$\exp(0,f(<..>)) = \begin{cases} 0 & \text{means "no closing; node is inside the tree"} \\ n\neq 0 & \text{means "closing by the basic functional with} \\ & \text{number n; node is on the bar or outside the tree"} \end{cases}$$

b) In case exp(0,f(<..>))=0:

$$\exp(1,f(<..>)) = \begin{cases} 0 & \text{means "node with infinitary cs-ramification"} \\ 1 & \text{means "node with unary (T2)-branching"} \\ 2 & \text{means "node with binary application ramification"} \\ 3 & \text{means "node with binary (T3)-branching"} \\ \geq 4 & \text{means "node with binary (T4)-branching"} \end{cases}$$

c) exp(2,f(<..>)) = type number of the term given by the subtree above f(<..>) (type number/f = exp(2,f(<>)))

Besides this we still have to arrange the type connexions.

d) $T(z)$ = type number of the basic functional with number z (primitive recursive)

e) For infinitary cs-ramifications:

$T_0(\lambda z f(\langle y\rangle * z),f) :\equiv \text{type}/f \equiv (\text{type}/\lambda z f(\langle y\rangle * z))0$
(primitive recursive in f)

f) For unary (T2)-branching:

$T_1(\lambda z f(\langle 0\rangle * z),f) :\equiv$ type/f from type/$\lambda z f(\langle 0\rangle * z)$ according to (T2)
(primitive recursive in f)

g) For binary application ramification:

$T_2(\lambda z f(\langle 0\rangle * z),\lambda z f(\langle 1\rangle * z),f) :\equiv$ type/f from type/$\lambda z f(\langle 0\rangle * z)$, type/$\lambda z f(\langle 1\rangle * z)$ according to application
(primitive recursive in f)

h) Analogously for binary (T3)-, (T4)-branchings with T_3, T_4.

It is technically simpler when one always speaks of infinitely ramified trees; this is achieved here by reiterating the last subtree at a finite ramification.

Corresponding to the so defined term construction one has to take the reduction instructions for the computation.

The generation instructions in XIII for the $\mathfrak{M}$-elements can now be formulated as follows:

$A_1(f) :\equiv \bigwedge y T_0(\lambda z f(\langle y\rangle * z),f)$

$A_2(f) :\equiv T_1(\lambda z f(\langle 0\rangle * z),f) \wedge \bigwedge y>0\ \lambda z f(\langle y\rangle * z)=\lambda z f(\langle 0\rangle * z)$

$A_3(f)$, $A_4(f)$ establish analogously the type connections and the reiterations of the last subtree for binary application and (T3)-branchings.

$A_5(f) :\equiv T_4(\lambda z f(\langle 0\rangle * z),\lambda z f(\langle 1\rangle * z),f) \wedge \bigwedge y\geq 2\ \lambda z f(\langle y\rangle * z)=\lambda z f(\langle 1\rangle * z)$

$\mathfrak{A}_0(\mathfrak{B},f) :\equiv \bigvee x,y(f=\lambda z\ 2^{x'}5^{T(x')}y')$ (basic functionals)

$\mathfrak{A}_1(\mathfrak{B},f) :\equiv \exp(0,f0)=0 \wedge \exp(1,f0)=0 \wedge \bigwedge y\ \lambda z f(\langle y\rangle * z)\in\mathfrak{B} \wedge A_1(f)$
(cs-ramification)

$\mathfrak{A}_2(\mathfrak{B},f) :\equiv \exp(0,f0)=0 \wedge \exp(1,f0)=1 \wedge \lambda z f(\langle 0\rangle * z)\in\mathfrak{B} \wedge A_2(f)$
((T2)-branching)

Analogously $\mathfrak{A}_3(\mathfrak{B},f)$, $\mathfrak{A}_4(\mathfrak{B},f)$ for binary application and (T3)-ramifications.

$\mathfrak{A}_5(\mathfrak{L},f) \equiv: \exp(0,f0)=0 \wedge \exp(1,f0)\geq 4 \wedge \lambda z f(\langle 0\rangle * z)\varepsilon\mathfrak{L} \wedge$
$\lambda z f(\langle 1\rangle * z)\varepsilon\mathfrak{L} \wedge A_5(f)$ ((T4)-branching)

$\mathfrak{A}(\mathfrak{L},f) \equiv: \mathfrak{A}_0(\mathfrak{L},f) \vee \ldots \vee \mathfrak{A}_5(\mathfrak{L},f)$

Immediately: $\mathfrak{L}_1 \subseteq \mathfrak{L}_2 \rightarrow \mathfrak{A}(\mathfrak{L}_1,f) \rightarrow \mathfrak{A}(\mathfrak{L}_2,f)$

The membership of $\mathfrak{M}$ itself can be expressed as follows:

Bar properties:
$G_{\mathfrak{M}}(f) \equiv: \bigwedge\alpha\bigvee y\{\exp(0,f(\overline{\alpha,y}))\neq 0 \wedge \exp(2,f(\overline{\alpha,y}))=T(\exp(0,f(\overline{\alpha,y})))\}$
$\wedge \bigwedge m,n\{\exp(0,fm)\neq 0 \rightarrow f(m*n)=fm\}$

Type connections and branching completions:
$T_{\mathfrak{M}}(f) \equiv: \bigwedge\alpha,x\{\exp(0,f(\overline{\alpha,x}))=0 \rightarrow (\exp(1,f(\overline{\alpha,x}))=0 \rightarrow A_1(f)) \wedge \ldots$
$\ldots \wedge (\exp(1,f(\overline{\alpha,x}))\geq 4 \rightarrow A_5(f))\}$

$f\varepsilon\mathfrak{M} \equiv: G_{\mathfrak{M}}(f) \wedge T_{\mathfrak{M}}(f)$

Now we have to show in the language $\mathcal{T}$:

$(*_1)$ $\mathfrak{A}(\mathfrak{M},f) \rightarrow f\varepsilon\mathfrak{M}$ is provable in Heyting arithmetic

$(*_2)$ $\bigwedge g\{\mathfrak{A}(\mathfrak{N},g) \rightarrow g\varepsilon\mathfrak{N}\} \wedge f\varepsilon\mathfrak{M} \rightarrow f\varepsilon\mathfrak{N}$ is provable in Heyting arithmetic plus BI_D^0

Remark:
Of course here one always has to read ε as an in $\mathcal{T}$ expressible predicate over the element!

Ad $(*_1)$:

$\mathfrak{A}_0(\mathfrak{M},f) \rightarrow f\varepsilon\mathfrak{M}$

For i=1,...,5: $\mathfrak{A}_i(\mathfrak{M},f) \rightarrow T_{\mathfrak{M}}(f)$

$\mathfrak{A}_i(\mathfrak{M},f) \rightarrow \bigwedge y\lambda z f(\langle y\rangle * z)\varepsilon\mathfrak{M} \wedge \exp(0,f0)=0$
$\rightarrow \bigwedge y,\beta\bigvee z\{\exp(0,f(\langle y\rangle * \overline{\beta,z}))\neq 0 \wedge \exp(2,f(\langle y\rangle * \overline{\beta,z}))=$
$T(\exp(0,f(\langle y\rangle * \overline{\beta,z})))\} \wedge \bigwedge y,m,n\{\exp(0,f(\langle y\rangle * m))\neq 0 \rightarrow$
$f(\langle y\rangle * m * n)=f(\langle y\rangle * m)\} \wedge \exp(0,f0)=0$
$\rightarrow \bigwedge\alpha\bigvee y\{\exp(0,f(\overline{\alpha,y}))\neq 0 \wedge \exp(2,f(\overline{\alpha,y}))=T(\exp(0,f(\overline{\alpha,y})))\}$
$\wedge \bigwedge m,n\{\exp(0,fm)\neq 0 \rightarrow f(m*n)=fm\}$
$\rightarrow f\varepsilon\mathfrak{M}$

Altogether: $\mathfrak{A}(\mathfrak{M},f) \rightarrow f\varepsilon\mathfrak{M}$

<u>Ad ($*_2$):</u>

(1) $V \equiv: \bigwedge g\{\mathfrak{A}(\mathfrak{N},g) \rightarrow g \varepsilon \mathfrak{N}\} \wedge f \varepsilon \mathfrak{M}$

(2) $A(\overline{\alpha,x}) \equiv: \exp(0,f(\overline{\alpha,x})) \neq 0 \wedge \exp(2,f(\overline{\alpha,x})) = T(\exp(0,f(\overline{\alpha,x})))$

(3) $B(\overline{\alpha,x}) \equiv: \lambda z f(\overline{\alpha,x*z}) \varepsilon \mathfrak{N}$

For this holds:

(4) $V \longrightarrow \bigwedge \alpha \bigvee y A(\overline{\alpha,y})$ (1)

(5) $V \longrightarrow \bigwedge \alpha\ ,x\{A(\overline{\alpha,x}) \vee \neg A(\overline{\alpha,x})\}$ (2)

(6) $V \wedge A(\overline{\alpha,x}) \longrightarrow \exp(0,f(\overline{\alpha,x})) \neq 0 \wedge f(\overline{\alpha,x*z}) = f(\overline{\alpha,x})$

$\wedge \exp(2,f(\overline{\alpha,x})) = T(\exp(0,f(\overline{\alpha,x})))$ (2),(1)

$\longrightarrow \bigvee x,y(\lambda z f(\overline{\alpha,x*z}) = \lambda z 2^{x'} 5^{T(x')} y')$

$\longrightarrow \mathfrak{A}_o(\mathfrak{N},\lambda z f(\overline{\alpha,x*z}))$

$\longrightarrow \mathfrak{A}(\mathfrak{N},\lambda z f(\overline{\alpha,x*z}))$

$\longrightarrow \lambda z f(\overline{\alpha,x*z}) \varepsilon \mathfrak{N}$ (1)

$\longrightarrow B(\overline{\alpha,x})$ (3)

(7) $V \wedge \exp(0,f(\overline{\alpha,x})) \neq 0 \longrightarrow f(\overline{\alpha,x*z}) = f(\overline{\alpha,x})$ (1)

$\longrightarrow A(\overline{\alpha,x})$ (2),(1)

$\longrightarrow B(\overline{\alpha,x})$ (6)

For $i=1,\ldots,5$ (with $\geq$ only in the case $i=5$):

(8) $V \wedge \bigwedge a^{o} B(\overline{\alpha,x*<a>}) \wedge \exp(0,f(\overline{\alpha,x})) = 0 \wedge \exp(1,f(\overline{\alpha,x})) \geqq i-1$

$\longrightarrow A_i(f) \wedge \bigwedge y\ \lambda z f(\overline{\alpha,x*<y>*z}) \varepsilon \mathfrak{N}$ (1),(3)

$\longrightarrow \mathfrak{A}_i(\mathfrak{N},\lambda z f(\overline{\alpha,x*z}))$

$\longrightarrow \mathfrak{A}(\mathfrak{N},\lambda z f(\overline{\alpha,x*z}))$

$\longrightarrow \lambda z f(\overline{\alpha,x*z}) \varepsilon \mathfrak{N}$ (1)

$\longrightarrow B(\overline{\alpha,x})$ (3)

From (7), (8):

(9) $V \longrightarrow \bigwedge \alpha,x\{\bigwedge a^{o} B(\overline{\alpha,x*<a>}) \longrightarrow B(\overline{\alpha,x})\}$

Now from (4), (5), (6), (9) by BI_D^o: $V \rightarrow B(<>)$, i.e. $V \rightarrow f \varepsilon \mathfrak{N}$.

Thus the model $\mathfrak{M}$ is completely characterized.

Bar induction over species is treated here in the general form $(aBI)_D$ where the species $S \equiv S(\alpha y, y; \overline{\alpha, y}) \equiv S(\overline{\alpha, y'}; y)$ may also depend on the path to the origin of the tree. The following considerations come off in the functional language and are based on <u>Heyting-arithmetic plus T, (ER)-qf.</u>

$$(H0)\ \bigwedge \alpha^{\sigma 0}, x^0 \{ \bigwedge y<x\ S(\overline{\alpha, y'}; y) \rightarrow \bigvee a^{\sigma}\ S(\overline{\alpha, x} * a; x)\}$$

$$(H1)\ \bigwedge \alpha^{\sigma 0} \{ \bigwedge y^0 S(\overline{\alpha, y'}; y) \rightarrow \bigvee z^0 A(\overline{\alpha, z}; z)\}$$

$$(H2)\ \bigwedge \alpha^{\sigma 0}, x^0 \{ \bigwedge y<x\ S(\overline{\alpha, y'}; y) \rightarrow A(\overline{\alpha, x}; x) \vee \neg A(\overline{\alpha, x}; x)\}$$

$$(H3)\ \bigwedge \alpha^{\sigma 0}, x^0 \{ \bigwedge y<x\ S(\overline{\alpha, y'}; y) \wedge A(\overline{\alpha, x}; x) \rightarrow B(\overline{\alpha, x}; x)\}$$

$$(H4)\ \bigwedge \alpha^{\sigma 0}, x^0 \{ \bigwedge y<x\ S(\overline{\alpha, y'}; y) \wedge \bigwedge a^{\sigma} (S(\overline{\alpha, x} * a; x) \rightarrow B(\overline{\alpha, x} * a; x')) \rightarrow B(\overline{\alpha, x}; x)\}$$

<u>General bar induction over species $(aBI)^{\sigma}_{D}$:</u>

$$(aBI)^{\sigma}_{D}\text{-rule: } \frac{(H1), \ldots, (H4)}{(H0) \rightarrow B(\sigma; 0)}$$

$$(aBI)^{\sigma}_{D}\text{: } (H0) \wedge \ldots \wedge (H4) \rightarrow B(\sigma; 0)$$

<u>Higher bar induction over species $(hBI)^{\sigma}_{D}$:</u>

Here the species S depends only on the distance to the origin of the tree: $S \equiv S(\alpha y; y)$.

$(hBI)^{\sigma}_{D}$-rule, $(hBI)^{\sigma}_{D}$ are the special cases of $(aBI)^{\sigma}_{D}$-rule, $(aBI)^{\sigma}_{D}$ for such species $S(\alpha y; y)$. (H0) here simplifies to $(H0)_h$: $\bigwedge x^0 \bigvee a^{\sigma}\ S(a; x)$.

<u>Higher bar induction over particularly one species $(hsBI)^{\sigma}_{D}$:</u>

Now only one species $S \equiv S(\alpha y)$ is considered.

$(hsBI)^{\sigma}_{D}$-rule, $(hsBI)^{\sigma}_{D}$ are the subcases of $(aBI)^{\sigma}_{D}$-rule, $(aBI)^{\sigma}_{D}$ for such species $S(\alpha y)$. (H0) here simplifies to $(H0)_{hs}$: $\bigvee a^{\sigma}\ S(a)$.

<u>Bar induction $(BI)^{\sigma}_{D}$:</u>

Here we have no species at all.

$(BI)^{\sigma}_{D}$-rule, $(BI)^{\sigma}_{D}$ are the special cases of $(aBI)^{\sigma}_{D}$-rule, $(aBI)^{\sigma}_{D}$ for $S \equiv \curlyvee$. (H0) is now superfluous.

Remark:
Similarly as to (BI) also versions $(aBI)_M$, $(aBI)_G$ etc. can be considered.

In the following the above notions will be analysed to a certain extent.

(III)

(a) $(aBI)_D^\tau \vdash (hBI)_D^\tau \vdash (hsBI)_D^\tau \vdash (BI)_D^\tau$

(b) $(aBI)_D^{\sigma 0} \vdash (aBI)_D^\sigma$

(c) $(hBI)_D^{\sigma 0} \vdash (hBI)_D^\sigma$

(d) $(hsBI)_D^{\sigma 0} \vdash (hsBI)_D^\sigma$

(e) $(BI)_D^{\sigma 0} \vdash (BI)_D^\sigma$

(f) The same deductive relations (a) - (e) also hold between the corresponding rules.

(g) (b) - (f) are generally valid for σ comprising types τ instead of $\sigma 0$.

Proof:
(a) and the corresponding (f)-part follow directly from the definitions.

Ad (b), (c) and corresponding (f)-part:
If one replaces in (H0) - (H4) $\alpha^{\sigma 0}$, a^σ by $\lambda u^0(\beta^{\sigma 00}u0)$, $b^{\sigma 0}0$ and quantifies accordingly to $\beta^{\sigma 00}$, $b^{\sigma 0}$, so one obtains:

$$(H0) \longrightarrow \begin{cases} \bigwedge \beta^{\sigma 00},x\{\bigwedge y<x \; S(\overline{\lambda u(\beta u0),y'};y) \longrightarrow \bigvee b^{\sigma 0} \; S(\overline{\lambda u(\beta u0),x*b0};x)\} \\ \bigwedge x \bigvee b^{\sigma 0} \; S(b0;x) \end{cases}$$

namely $b=\lambda u^0 a$

$$(H1) \longrightarrow \bigwedge \beta^{\sigma 00}\{\bigwedge y \begin{cases} S(\overline{\lambda u(\beta u0),y'};y) \\ S(\beta y0;y) \end{cases} \longrightarrow \bigvee z A(\overline{\lambda u(\beta u0),z};z)\}$$

$$(H2) \longrightarrow \bigwedge \beta^{\sigma 00},x\{\bigwedge y<x \begin{cases} S(\overline{\lambda u(\beta u0),y'};y) \\ S(\beta y0;y) \end{cases} \longrightarrow A(\overline{\lambda u(\beta u0),x};x) \vee \neg A(\overline{\lambda u(\beta u0),x};x)\}$$

$$(H3) \longrightarrow \bigwedge \beta^{\sigma oo},x\{ \bigwedge y<x \left\{\begin{matrix} S(\overline{\lambda u(\beta u0),y'};y) \\ \\ S(\beta y0;y) \end{matrix}\right. \wedge A(\overline{\lambda u(\beta u0),x};x) \longrightarrow B(\overline{\lambda u(\beta u0),x};x)\}$$

$$(H4) \longrightarrow \bigwedge \beta^{\sigma oo},x\{ \bigwedge y<x \left\{\begin{matrix} S(\overline{\lambda u(\beta u0),y'};y) \\ \\ S(\beta y0;y) \end{matrix}\right. \wedge \bigwedge b^{\sigma o}(\left\{\begin{matrix} S(\overline{\lambda u(\beta u0),x}*b0;x) \\ \\ S(b0;x) \end{matrix}\right. \longrightarrow B(\overline{\lambda u(\beta u0),x}*b0;x')) \longrightarrow B(\overline{\lambda u(\beta u0),x};x)\}$$

The relations

$$\overline{\lambda u^{o}(\beta^{\sigma oo}u0),x^{\sigma}} = \lambda u^{o}(\overline{\beta,x}\ u0)$$

$\overline{\lambda u(\beta u0),x*b^{\sigma o}0} = \lambda u((\overline{\beta,x}*b)u0)$ resulting from (ER)-qf further yield:

$$(H0) \longrightarrow \left\{\begin{matrix} \bigwedge \beta^{\sigma oo},x\{ \bigwedge y<x\ S(\lambda u(\overline{\beta,y'}u0);y) \longrightarrow \bigvee b^{\sigma o}\ S(\lambda u((\overline{\beta,x}*b)u0);x)\} \\ \\ \bigwedge x \bigvee b^{\sigma o}\ S(b0;x) \end{matrix}\right.$$

$$(H1) \longrightarrow \bigwedge \beta^{\sigma oo}\{ \bigwedge y \left\{\begin{matrix} S(\lambda u(\overline{\beta,y'}u0);y) \\ \\ S(\beta y0;y) \end{matrix}\right. \longrightarrow \bigvee z A(\lambda u(\overline{\beta,z}u0);z)\}$$

$$(H2) \longrightarrow \bigwedge \beta^{\sigma oo},x\{ \bigwedge y<x \left\{\begin{matrix} S(\lambda u(\overline{\beta,y'}u0);y) \\ \\ S(\beta y0;y) \end{matrix}\right. \longrightarrow A(\lambda u(\overline{\beta,x}u0);x) \vee \neg A(\lambda u(\overline{\beta,x}u0);x)\}$$

$$(H3) \longrightarrow \bigwedge \beta^{\sigma oo},x\{ \bigwedge y<x \left\{\begin{matrix} S(\lambda u(\overline{\beta,y'}u0);y) \\ \\ S(\beta y0;y) \end{matrix}\right. \wedge A(\lambda u(\overline{\beta,x}u0);x) \longrightarrow B(\lambda u(\overline{\beta,x}u0);x)\}$$

$$(H4) \longrightarrow \bigwedge \beta^{\sigma oo},x\{ \bigwedge y<x \left\{\begin{matrix} S(\lambda u(\overline{\beta,y'}u0);y) \\ \\ S(\beta y0;y) \end{matrix}\right. \wedge \bigwedge b^{\sigma o}(\left\{\begin{matrix} S(\lambda u((\overline{\beta,x}*b)u0);x) \\ \\ S(b0;x) \end{matrix}\right. \longrightarrow B(\lambda u((\overline{\beta,x}*b)u0);x')) \longrightarrow B(\lambda u(\overline{\beta,x}u0),u)\}$$

From this now by $(aBI)^{\sigma o}_{D}$ resp. $(hBI)^{\sigma o}_{D}$

$(H0) \wedge \ldots \wedge (H4) \longrightarrow B(\lambda u(\sigma u0);0)$

and with (ER)-qf finally $(aBI)^{\sigma}_{D}$ resp. $(hBI)^{\sigma}_{D}$.

The derivability of the corresponding rules comes off analogously.

Ad (d), (e) and corresponding (f)-part:

These proofs turn out as the special cases $S \equiv S(\alpha y)$ resp. $S\equiv Y$ of the above demonstration for (c).

Ad (g):

Actually the above proof can be generalized to a derivability for type σ from comprising types τ by means of an explicit functional operation $F^{\sigma\tau}$ with

(1) $\bigwedge a^{\sigma} \bigvee b^{\tau}\ Fb=a$ (2) $F\sigma=\sigma$

(Above $F=\lambda b^{\sigma o} b0$ was used.) For now one replaces in (H0) - (H4) $\alpha^{\sigma o}$, a^{σ} by $\lambda u^{o} F(\beta^{\tau o} u)$, Fb^{τ} and obtains with (1) the first five implications. From (2) follows extensionally:

$$\overline{\lambda u^{o} F(\beta u), x^{o}} = \lambda u F(\overline{\beta^{\tau o}, x}\ u)$$

$$\overline{\lambda u F(\beta u), x} * (Fb) = \lambda u F((\overline{\beta, x} * b^{\tau}) u)$$

By this one gets the corresponding further implications. With the assumed bar induction resp. bar induction rule of type τ using (2) the desired conclusion now follows.

For instance in the case $\sigma \equiv :00$, $\tau \equiv :0(00)$ one can take the functional operation $F^{\sigma\tau} b =: \lambda u^{o} b(\lambda z^{o} u)$ ((1) is satisfied by $b^{\tau} =: \lambda f^{oo} a^{\sigma}(f0)$.).

(IV)

$(aBI)^{\sigma}_{D}$-rule $\vdash (aBI)^{\sigma}_{D}$

$(hBI)^{\sigma}_{D}$-rule $\vdash (hBI)^{\sigma}_{D}$

$(hsBI)^{\sigma}_{D}$-rule $\vdash (hsBI)^{\sigma}_{D}$

Proof: The derivability is shown simultaneously for the general and higher bar induction over species; the $(hsBI)_D$-case follows from the $(hBI)_D$-case by cancellation of the type 0-argument in S, S_1 and the premise $y>0$ in S_1.

$$S_1\left(\begin{matrix} v^{\sigma o} \\ \\ v^{\sigma} \end{matrix} ; y^{o}\right) \equiv: (H1) \wedge (H2) \wedge \left(y>0 \longrightarrow S\left(\begin{matrix} \lambda u^{o} v^{\sigma o} u' \\ \\ v^{\sigma} \end{matrix} ; \delta y\right)\right)$$

$$A_1(w^{\sigma o}; x^{o}) \equiv: x>0 \wedge A(\overline{\lambda u^{o}(wu'), \delta x}; \delta x)$$

$B_1(w^{\sigma o};x^o) :\equiv (H0) \wedge \dots \wedge (H4) \longrightarrow B(\overline{\lambda u^o(wu'),\delta x};\delta x)$

with predecessor function δ

By (ER)-qf one proves the following relations:

(1) $\overline{\lambda u^o(\overline{\alpha,x})u',\delta x} = \overline{\lambda u\alpha u',\delta x}$

(2) $\overline{\lambda u^o(\overline{\alpha,x*a})u',x^\sigma} = \overline{\overline{\lambda u^o\alpha u',\delta x*a},x}$

(3) $x>0 \longrightarrow \overline{\overline{\gamma,\delta x*a},x} = \overline{\gamma,\delta x*a}$

(4) $\lambda u^o(\overline{\alpha,y'u'}) = \overline{\lambda u\alpha u',y}$

(5) $x>0 \longrightarrow \lambda u(\overline{\alpha,x*a})u' = \overline{\lambda u\alpha u',\delta x*a}$

The premises of the $(aBI)^\sigma_D$- resp. $(hBI)^\sigma_D$-rule with S_1, A_1, B_1 can now be proved as follows.

(6) $\bigwedge y^o S_1\left(\begin{matrix}\overline{\alpha,y'}\\ \\ \alpha y\end{matrix};y\right) \longrightarrow \bigvee z^o A_1(\overline{\alpha,z};z)$

<u>Proof:</u> Because of (1), (4) it suffices to show:

$$\bigwedge\beta^{\sigma o}\{\bigwedge y^o S\left(\begin{matrix}\overline{\beta,y'}\\ \\ \beta y\end{matrix};y\right) \longrightarrow \bigvee x^o A(\overline{\beta,x};x)\} \wedge \bigwedge y>0\ S\left(\begin{matrix}\overline{\lambda u\alpha u',y}\\ \\ \alpha y\end{matrix};\delta y\right) \longrightarrow$$
$$\bigvee z\{z>0 \wedge A(\overline{\lambda u\alpha u',\delta z};\delta z)\}$$

The conclusion follows from the premises for $\beta :\equiv \lambda u^o\alpha u'$, $z :\equiv x'$.

(7) $\bigwedge y<x^o S_1\left(\begin{matrix}\overline{\alpha,y'}\\ \\ \alpha y\end{matrix};y\right) \longrightarrow A_1(\overline{\alpha,x};x) \vee \neg A_1(\overline{\alpha,x};x)$

<u>Proof:</u> For x=0 in the conclusion the second member of the disjunction becomes true; in case x>0 on account of (1),(4) we have to show:

$$\bigwedge\beta^{\sigma o},z^o\{\bigwedge y<z\ S\left(\begin{matrix}\overline{\beta,y'}\\ \\ \beta y\end{matrix};y\right) \longrightarrow A(\overline{\beta,z};z) \vee \neg A(\overline{\beta,z};z)\} \wedge \bigwedge y<x(y>0 \longrightarrow$$
$$S\left(\begin{matrix}\overline{\lambda u\alpha u',y}\\ \\ \alpha y\end{matrix};\delta y\right)) \longrightarrow A(\overline{\lambda u\alpha u',\delta x};\delta x) \vee \neg A(\overline{\lambda u\alpha u',\delta x};\delta x)$$

Here the conclusion follows from the premises with $\beta :\equiv \lambda u\alpha u'$, $z :\equiv \delta x < x$.

$$(8)\quad \bigwedge y{<}x^{o} S_1\left(\left\{\begin{array}{l}\overline{\alpha\,,y'}\\ \\ \alpha\,y\end{array}\right.;y\right) \wedge A_1(\overline{\alpha\,,x};x) \longrightarrow B_1(\overline{\alpha\,,x};x)$$

<u>Proof:</u> For x=0 the second premise becomes false; in case x>0 on account of (1), (4) we have to show:

$$\bigwedge y{<}x(y{>}0 \rightarrow S\left(\left\{\begin{array}{l}\overline{\lambda u\alpha u',y}\\ \\ \alpha y\end{array}\right.;\delta y)\right) \wedge A(\overline{\lambda u\alpha u',\delta x};\delta x) \wedge \bigwedge \beta,z\{\bigwedge y{<}z S\left(\left\{\begin{array}{l}\overline{\beta,y'}\\ \\ \beta y\end{array}\right.;y\right)\wedge$$

$$A(\overline{\beta,z};z) \longrightarrow B(\overline{\beta,z};z)\} \longrightarrow B(\overline{\lambda u\alpha u',\delta x};\delta x)$$

The conclusion follows from the premises for $\beta \equiv : \lambda u\alpha u'$, $z \equiv : \delta x < x$.

$$(9)\quad \bigwedge y{<}x^{o} S_1\left(\left\{\begin{array}{l}\overline{\alpha,y'}\\ \\ \alpha y\end{array}\right.;y\right) \wedge \bigwedge a(S_1\left(\left\{\begin{array}{l}\overline{\alpha,x{*}a}\\ \\ \quad a\end{array}\right.;x\right) \rightarrow B_1(\overline{\alpha,x{*}a};x')) \longrightarrow B_1(\overline{\alpha,x};x)$$

<u>Proof:</u> Because of (1) - (5) it remains to show:

$$\bigwedge y{<}x(y{>}0 \rightarrow S\left(\left\{\begin{array}{l}\overline{\lambda u\alpha u',y}\\ \\ \alpha y\end{array}\right.;\delta y)\right) \wedge \bigwedge a((x{>}0 \rightarrow S\left(\left\{\begin{array}{l}\overline{\lambda u\alpha u',\delta x{*}a}\\ \\ \quad a\end{array}\right.;\delta x)\right) \longrightarrow$$

$$B(\overline{\lambda u\alpha u',\delta x{*}a,x};x)) \wedge (HO) \wedge \bigwedge \beta,z\{\bigwedge y{<}z S\left(\left\{\begin{array}{l}\overline{\beta,y'}\\ \\ \beta y\end{array}\right.;y\right)\wedge$$

$$\bigwedge a(S\left(\left\{\begin{array}{l}\overline{\beta,z{*}a}\\ \\ \quad a\end{array}\right.;z\right) \rightarrow B(\overline{\beta,z{*}a};z')) \rightarrow B(\overline{\beta,z};z)\} \longrightarrow$$

$$B(\overline{\lambda u\alpha u',\delta x};\delta x)$$

For $0=x=\delta x$ the conclusion follows with (ER)-qf from the second premise. However in the proof of the $(hsBI)_D$-case x>0 falls away in the second premise; here now the further assumption $(HO)_{hs}$: $\bigvee aS(a)$ is used.

In case x>0 the conclusion follows from the premises 1,2 and 4 for $\beta \equiv : \lambda u\alpha u'$, $z \equiv : \delta x < x = z'$ with (3).

From (6) - (9) now the $(aBI)^{\sigma}_{D}$- resp. $(hBI)^{\sigma}_{D}$-rule yields $(HO) \longrightarrow B_1(\sigma;0)$; from this with (ER)-qf $(aBI)^{\sigma}_{D}$ resp. $(hBI)^{\sigma}_{D}$.

<u>Addition:</u> Because (HO) is used in the preceding proof only in the $(hsBI)_D$-case, so (IV) also holds for general and higher bar induction over species when (HO) is omitted.

We still mention that like bar induction over species also $(MP)^\sigma$ is deductive equivalent to the $(MP)^\sigma$-rule:

$$\frac{\bigwedge x^\sigma (A(x) \vee \neg A(x)) \longrightarrow \neg\neg \bigvee x A(x)}{\bigwedge x (A(x) \vee \neg A(x)) \longrightarrow \bigvee x A(x)}$$

(V) $(MP)^\sigma$-rule $\vdash (MP)^\sigma$

Proof: $B(u^\sigma) :\equiv: \neg\neg\bigvee x A(x) \longrightarrow A(u)$

$$\begin{aligned} \bigwedge x(B(x) \vee \neg B(x)) &\longrightarrow \neg\neg\bigvee y B(y) \\ &\longrightarrow \bigvee y B(y) && (MP)^\sigma\text{-rule} \\ &\longrightarrow \neg\neg\bigvee x A(x) \longrightarrow \bigvee y A(y) \end{aligned}$$

With $\neg\neg\bigvee x A(x) \longrightarrow \{B(u) \longleftrightarrow A(u)\}$ now $(MP)^\sigma$.

Characteristic of our proceeding is the following compactness.

(VI) $(aBI)^\sigma_D$, $(hBI)^\sigma_D$, $(hsBI)^\sigma_D$ have a functional interpretation in the narrower sense in $T \cup BR$.

This follows by (III) from the deductive connection:

(VII) Heyting-analysis plus $(MP), (\underset{\rightarrow}{\bigvee}), (BI)^\alpha_M \vdash (aBI)^\sigma_D$ without (HO)

Because a functional interpretation in the narrower sense was given by (6.4) for Heyting-analysis plus $(MP), (\underset{\rightarrow}{\bigvee})$ in T and for $(BI)^\alpha_M$ in $T \cup BR^\alpha$ by Howard [18].

With (VII) one also has a classical proof for $(aBI)^\sigma_D$, $(hBI)^\sigma_D$, $(hsBI)^\sigma_D$ without (HO), (H2) (Howard-Kreisel [19], 351-353) what however can already be done directly in (ωAC)-analysis in a much shorter way:

$$\begin{aligned}
&\text{(H1)} \longleftrightarrow \bigwedge \alpha^{\sigma o}\{\bigvee y \neg S(\overline{\alpha, y'}; y) \vee \bigvee z A(\overline{\alpha, z}; z)\} \\
\text{(1)} \quad &\phantom{\text{(H1)}} \longleftrightarrow \bigwedge \alpha \bigvee z\{\bigwedge y{<}z S(\overline{\overline{\alpha, z}, y'}; y) \longrightarrow A(\overline{\alpha, z}; z)\} \\
\text{(2)} \quad &\text{(H3)} \longleftrightarrow \bigwedge \alpha, x\{(\bigwedge y{<}x S(\overline{\overline{\alpha, x}, y'}; y) \longrightarrow A(\overline{\alpha, x}; x)) \longrightarrow \\
&\qquad\qquad (\bigwedge y{<}x S(\overline{\overline{\alpha, x}, y'}; y) \longrightarrow B(\overline{\alpha, x}; x))\} \\
\text{(3)} \quad &\text{(H4)} \longleftrightarrow \bigwedge \alpha, x\{\bigwedge a(\bigwedge y{<}x' S(\overline{\overline{\alpha, x{*}a}, y'}; y) \longrightarrow B(\overline{\alpha, x{*}a}; x')) \longrightarrow \\
&\qquad\qquad (\bigwedge y{<}x S(\overline{\overline{\alpha, x}, y'}; y) \longrightarrow B(\overline{\alpha, x}; x))\}
\end{aligned}$$

From (1) - (3) with $(BI)_D^\sigma$: $(H1) \wedge (H3) \wedge (H4) \longrightarrow \bigwedge y{<}0\ S \longrightarrow B(\sigma;0)$
$\longrightarrow B(\sigma;0)$

Thus classically the above considered bar inductions over species of type σ - where (H0), (H2) may also be omitted - are not stronger than $(BI)_D^\sigma$.

Proof of (VII):

After (4.2): (1) $S(u^{\sigma o};y^o) \longleftrightarrow \bigvee v^\tau \bigwedge w S_o(u;y;v;w)$
with quantifierfree S_o

With (1) from (H1) - (H4) by (AC), $(\xrightarrow{V})$ now follows:

(2) $(H1) \longleftrightarrow \bigwedge \alpha^{\sigma o}, V^{\tau o} \bigvee z^o\{ \bigwedge y^o, w S_o(\overline{\alpha,y'};y;Vy;w) \longrightarrow A(\overline{\alpha,z};z)\}$

(3) $(H2) \longleftrightarrow \bigwedge \alpha^{\sigma o}, V^{\tau o}, x^o\{ \bigwedge y{<}x, \bigwedge w S_o(\overline{\alpha,y'};y;Vy;w) \longrightarrow A(\overline{\alpha,x};x) \vee \neg A(\overline{\alpha,x};x)\}$

(4) $(H3) \longleftrightarrow \bigwedge \alpha, V, x\{ \bigwedge y{<}x, \bigwedge w S_o(\overline{\alpha,y'};y;Vy;w) \wedge A(\overline{\alpha,x};x) \longrightarrow B(\overline{\alpha,x};x)\}$

(5) $(H4) \longleftrightarrow \bigwedge \alpha, V, x\{ \bigwedge y{<}x, \bigwedge w S_o(\overline{\alpha,y'};y;Vy;w) \wedge \bigwedge a^\sigma, v^\tau (\bigwedge w S_o(\overline{\alpha,x}*a;$
$x;v;w) \longrightarrow B(\overline{\alpha,x}*a;x')) \longrightarrow B(\overline{\alpha,x};x)\}$

$(\bigwedge y^o, w S_o(\overline{\alpha,y'};y;Vy;w) \longrightarrow A(\overline{\alpha,x};x)) \wedge (H2) \wedge \bigwedge y{<}x^o, \bigwedge w S_o(\overline{\alpha,y'};y;Vy;w)$
$\longrightarrow A(\overline{\alpha,x};x) \vee \neg A(\overline{\alpha,x};x)$ (3)
$\longrightarrow A(\overline{\alpha,x};x) \vee \neg \bigwedge y^o, w S_o(\overline{\alpha,y'};y;Vy;w)$
$\longrightarrow A(\overline{\alpha,x};x) \vee \bigvee y^o \neg \bigwedge w S_o(\overline{\alpha,y'};y;Vy;w)$ (MP)
$\longrightarrow \bigvee z^o\{ \bigwedge y{<}z, \bigwedge w S_o(\overline{\alpha,y'};y;Vy;w) \longrightarrow A(\overline{\alpha,z};z)\}$

From this with (2), $(\xrightarrow{V})$:

$(H1) \wedge (H2) \longrightarrow \bigvee x^o, z^o\{ \bigwedge y{<}x, \bigwedge w S_o(\overline{\alpha,y'};y;Vy;w) \wedge \bigwedge y{<}z, \bigwedge w S_o(\overline{\alpha,y'};y;Vy;w)$
$\longrightarrow A(\overline{\alpha,z};z)\}$

For k = max(x,z)+1:

$(H1) \wedge (H2) \longrightarrow \bigvee k^o\{ \bigwedge y{<}k, \bigwedge w S_o(\overline{\alpha,y'};y;Vy;w) \longrightarrow \bigvee y{<}k\ A(\overline{\alpha,y};y)\}$

(6) $\longrightarrow \bigvee k^o\{ \bigwedge y{<}k, \bigwedge w S_o(\overline{\overline{\alpha,k},y'};y;\overline{V,k}y;w) \longrightarrow \bigvee y{<}k\ A(\overline{\overline{\alpha,k},y};y)\}$

With the abbreviations

(7) $A_1(u^{\sigma o};k^o) \equiv: \bigvee y{<}k\ A(\overline{u,y};y)$

(8) $P(u^{\sigma o};v^{\tau o};k) \equiv: \bigwedge y{<}k, \bigwedge w S_o(\overline{u,y'};y;vy;w) \longrightarrow A_1(u;k)$

it is thus proved in (6):

(9) $(H1) \wedge (H2) \longrightarrow \bigwedge \alpha^{\sigma o}, V^{\tau o} \bigvee k^{o}\ P(\overline{\alpha,k};\overline{V,k};k)$

From (7) follows directly

$$k<l \wedge A_1(\overline{\alpha,k};k) \longrightarrow A_1(\overline{\alpha,l};l)$$

Now with (8)

$$(10)\quad \bigwedge \alpha^{\sigma o}, V^{\tau o}, a^{\sigma}, v^{\tau}, k^{o}\{P(\overline{\alpha,k};\overline{V,k};k) \longrightarrow P(\overline{\alpha,k{*}a};\overline{V,k{*}v};k')\}$$

Further abbreviations:

$$(11)\quad B_1(u^{\sigma o};k^{o}) \ \equiv:\ A_1(u;k) \vee B(u;k)$$

$$(12)\quad Q(u^{\sigma o};v^{\tau o};k^{o}) \ \equiv:\ \bigwedge y<k, \bigwedge w S_o(\overline{u,y'};y;vy;w) \longrightarrow B_1(u;k)$$

(8), (11), (12) immediately yield

$$(13)\quad \bigwedge \alpha^{\sigma o}, V^{\tau o}, k^{o}\{P(\overline{\alpha,k};\overline{V,k};k) \longrightarrow Q(\overline{\alpha,k};\overline{V,k};k)\}$$

$$\begin{aligned}
(H2) \wedge (H3) \wedge \bigwedge y<x, \bigwedge w S_o(\overline{\alpha,y'};y;Vy;w) &\longrightarrow \bigwedge l\leq x\{A(\overline{\alpha,l};l) \vee \neg A(\overline{\alpha,l};l)\} && (3)\\
&\longrightarrow \bigvee l\leq x A(\overline{\alpha,l};l) \vee \neg \bigvee l\leq x A(\overline{\alpha,l};l) && \text{(Kleene [17], 191, *150)}\\
&\longrightarrow \bigvee l<x A(\overline{\overline{\alpha,x},l};l) \vee A(\overline{\alpha,x};x) \vee \bigwedge a^{\sigma}\neg \bigvee l<x' A(\overline{\overline{\alpha,x{*}a},l};l) &&\\
&\longrightarrow A_1(\overline{\alpha,x};x) \vee B(\overline{\alpha,x};x) \vee \bigwedge a^{\sigma}\neg A_1(\overline{\alpha,x{*}a};x') && (7),\ (4)\\
&\longrightarrow B_1(\overline{\alpha,x};x) \vee \bigwedge a^{\sigma}\neg A_1(\overline{\alpha,x{*}a};x') && (11)
\end{aligned}$$

From this one obtains:

$$\begin{aligned}
&(H2) \wedge (H3) \wedge (H4) \wedge \bigwedge y<x, \bigwedge w S_o(\overline{\alpha,y'};y;Vy;w) \wedge \bigwedge a^{\sigma}, v^{\tau}(\bigwedge w S_o(\overline{\alpha,x{*}a};x;v;w)\\
&\quad \longrightarrow B_1(\overline{\alpha,x{*}a};x')) \longrightarrow B_1(\overline{\alpha,x};x) \vee \{\bigwedge a \neg A_1(\overline{\alpha,x{*}a};x') \wedge (H4) \wedge\\
&\qquad \bigwedge a^{\sigma}, v^{\tau}(\bigwedge w S_o(\overline{\alpha,x{*}a};x;v;w) \longrightarrow A_1(\overline{\alpha,x{*}a};x')\\
&\qquad \vee B(\overline{\alpha,x{*}a};x'))\} \qquad (11)\\
&\quad \longrightarrow B_1(\overline{\alpha,x};x) \qquad (5),\ (11)
\end{aligned}$$

Thus:

$$\begin{aligned}
&(H2) \wedge (H3) \wedge (H4) \longrightarrow \bigwedge a^{\sigma}, v^{\tau}(\bigwedge y<x', \bigwedge w S_o(\overline{(\overline{\alpha,x{*}a}),y'};y;(\overline{V,x{*}v})y;w) \longrightarrow\\
&\quad B_1(\overline{\alpha,x{*}a};x')) \longrightarrow \bigwedge y<x, \bigwedge w S_o(\overline{\overline{\alpha,x},y'};y;\overline{V,x}y;w) \longrightarrow\\
&\quad B_1(\overline{\alpha,x};x)
\end{aligned}$$

After (12) this means

$$(14)\quad (H2) \wedge (H3) \wedge (H4) \longrightarrow \bigwedge \alpha^{\sigma o}, V^{\tau o}, x^{o}\{\bigwedge a^{\sigma}, v^{\tau} Q(\overline{\alpha,x{*}a};\overline{V,x{*}v};x') \longrightarrow Q(\overline{\alpha,x};\overline{V,x};x)\}$$

From (9), (10), (13), (14) now by $(BI)_M$ after suitable contraction of α, V resp. a, v:

$(H1)\wedge \ldots \wedge (H4) \longrightarrow Q(\sigma;\sigma;0)$

$\longrightarrow \bigwedge y<0, \bigwedge wS_0 \longrightarrow \bigvee y<0A \vee B(\sigma;0)$ (12),(11),(7)

$\longrightarrow B(\sigma;0)$

Altogether to the following also classically interpretable domain of the functional language of finite types a functional interpretation in the narrower sense has been given in T∪BR:

(✱) Heyting-arithmetic plus T, (ER)-qf, (AC), (MP), $(\underset{\rightarrow}{V})$, $(\overset{\neg\neg}{\omega}AC)$, $(aBI)_D$

In the following we will consider deductive domains $\mathfrak{D}$ (within the functional language) which have a functional interpretation in the narrower sense in general recursive and consistently computable extensions I of T∪BR; always $(✱) \subseteq \mathfrak{D}$. Let $\mathfrak{D}_0$ correspond to T∪BR.

(VIII) $\mathfrak{D}$ plus $(C)^0$, in which at most type 0-parameters $\nvdash \lambda$

Proof: With (✱) the deduction theorem holds for $\mathfrak{D}$. Thus $\mathfrak{D}$ plus $(C)^0$ (type 0-parameters) $\vdash \lambda$ implies $\mathfrak{D} \vdash \neg\neg\bigwedge x_1^0(C)_1^0 \wedge \ldots \wedge \neg\neg\bigwedge x_n^0(C)_n^0 \longrightarrow \lambda$; with $\mathfrak{D} \vdash (✱) \vdash (\overset{\neg\neg}{\omega}AC) \vdash (\neg\bigwedge\neg)^0 \vdash \neg\neg\bigwedge x^0(Tnd) \vdash \neg\neg\bigwedge x^0(C)^0$ therefore $\mathfrak{D} \vdash \lambda$ contradicting the assumptions on $\mathfrak{D}$.

By restriction of $(C)^0$ in (VIII) to type 0-parameters on account of

$\bigwedge u^\alpha \bigvee z^{00} \bigwedge x^0\{zx=0 \longleftrightarrow A(u) \wedge x=x\}$ $(C)^0$ with type α-parameter

$\bigvee z^{00\alpha} \bigwedge u^\alpha, x^0\{zux=0 \longleftrightarrow A(u) \wedge x=x\}$ (AC)

$\bigvee z^{0\alpha} \bigwedge u^\alpha\{zu=0 \longleftrightarrow A(u)\}$

higher comprehensions are excluded. T∪BR is the hitherto largest functional interpretation domain; for $\alpha \neq 0$ we have $\mathfrak{D}_0 \nvdash \neg\neg(C)^\alpha$, $\neg\neg\bigwedge x^\alpha(Tnd)$, $(\neg\bigwedge\neg)^\alpha$ because otherwise the $(C)^\alpha$-analysis would have the same proof theoretic strength as the $(C)^0$-analysis.

(IX) (a) $\mathfrak{D} \vdash \neg((ChT)\text{-instance})$

(b) $\mathfrak{D}$ plus $(C)^0$, in which at most type 0-parameters $\nvdash \neg\neg(ChT)$

Proof: According to chapter IX,3: $\mathfrak{D} \vdash (✱) \vdash \neg((ChT)\text{-instance})$.
$\mathfrak{D}$ plus $(C)^0$(type α-parameters) $\vdash \neg\neg(ChT)$ therefore contradicts (VIII).

Because of (9.2) the conjecture $(aBI)_D,(ChT) \nvdash \bot$ cannot be decided by Gödel's functional interpretation.

(X) For subdomains $\mathfrak{T}_1$ of $\mathfrak{T}$ which contain besides Heyting-arithmetic also T, (ER)-qf, (AC), (MP), $(\underline{\vee}\!\!\rightarrow)$ and their functional interpretation domain $I_1 \subseteq I$ the following holds:

(a) $\vdash D \rightarrow A \vee B \Longleftrightarrow \vdash D \rightarrow A$ or $\vdash D \rightarrow B$

(b) $\vdash D \rightarrow \bigwedge x \bigvee y C(x,y) \Longleftrightarrow \vdash D \rightarrow \bigwedge x C(x,\varphi x)$ for a constant $\varphi \varepsilon I_1$

where A,B, $\bigwedge x \bigvee y C(x,y)$, D are closed and $D' \equiv \bigwedge z D_o(z)$ (D_o quantifierfree)

<u>Proof:</u> (a) reduces by $A \vee B \longleftrightarrow \bigvee t^o\{(t=0 \wedge A) \vee (t \neq 0 \wedge B)\}$ to (b).

Ad (b):

$\vdash D \rightarrow \bigwedge x \bigvee y C(x,y)$ gives a functional solution of

$$(\bigwedge z D_o(z) \rightarrow \bigwedge x \bigvee y \bigvee v \bigwedge w C_o(x,y,v,w))' \equiv \bigvee \varphi,\psi,\chi \bigwedge x,w\{D_o(\chi x w) \rightarrow C_o(x,\varphi x,\psi x,w)\}$$

in I_1. From this now:

$$\bigwedge z D_o(z) \rightarrow \bigwedge x \bigvee v \bigwedge w C_o(x,\varphi x,v,w)$$

$$D \rightarrow \bigwedge x C(x,\varphi x) \qquad (4.2)$$

Remark: In case (b) C,D,φ may also have parameters.

At the end of [40] Scarpellini mentions that an extension of his methods [39] gives corresponding results (X) for the formalization version (I) of our consistency proof. Certainly for all principles here treated the given intuitive motivation is confirmed by such intuitionistic formal results.

(XI) $\mathfrak{T} \nvdash (C)^\alpha$, $\bigwedge x^\alpha(\text{Tnd})$

<u>Proof:</u> Assume $\vdash$; then with (3.15), (*)

$\mathfrak{T} \vdash \bigwedge x^\alpha\{A(x) \vee \neg A(x)\}$ where $\alpha \equiv 0\alpha_1 \ldots \alpha_n$ $(n \geq 0)$

So specially for $x^\alpha :\equiv \lambda u_n^{\alpha_n} \ldots u_1^{\alpha_1} z^o$, $A(x) :\equiv \bigvee y^o T_1(x\sigma^{\alpha_n} \ldots \sigma^{\alpha_1} x\sigma \ldots \sigma, y)$ there would be a functional interpretation in the narrower sense of

(1) $\bigwedge z^o\{\bigvee y T_1(z,z,y) \vee \neg \bigvee y T_1(z,z,y)\}$

in a T∪BR-extension with general recursive computation instructions:

(1)' $\equiv \bigvee T^{oo}, X^{oo} \bigwedge z^o, y^o\{(Tz=0 \wedge T_1(z,z,Xz)) \vee (Tz \neq 0 \wedge \neg T_1(z,z,y))\}$

$\bigvee y T_1(z,z,y)$ would be recursively decidable by $Tz=0$. Contradiction.

From (XI) especially follows that in such domains $\mathcal{L}$ (respectively in therein interpretable other language and deduction connections) (C) is not provable from $(aBI)_D$. But also a converse independence can be shown.

(XII) Provided ZF-set theory is consistent so $ZF \nvdash (BI)_D^{\alpha}$ for $\alpha \not\equiv 0$

Proof:

Assumption: $ZF \vdash (BI)_D^{\alpha}$ for $\alpha \equiv 0\alpha_1 \ldots \alpha_n$ $(n \geq 1)$

$\vdash (AC)^{0\alpha}$ (Howard-Kreisel [19], 353)

(1) $\vdash (AC)^{01}$ because:

$$\bigwedge x^0 \bigvee f^{00} A(x,f) \longrightarrow \bigwedge x^0 \bigvee z^{\alpha} A(x, \lambda y^0 z\sigma^{\alpha_n} \ldots \sigma^{\alpha_2}(\lambda x_m^{\beta_m} \ldots x_1^{\beta_1}\, y))$$

where $\alpha_1 \equiv 0\beta_1 \ldots \beta_m$ $(m \geq 0)$, $z a_n^{\alpha_n} \ldots a_1^{\alpha_1} \stackrel{\circ}{=} f(a_1 \sigma^{\beta_m} \ldots \sigma^{\beta_1})$

$$\longrightarrow \bigvee z^{\alpha 0} \bigwedge x^0 A(x, \lambda y^0 z x \sigma \ldots \sigma(\lambda x_m \ldots x_1\, y)) \qquad (AC)^{0\alpha}$$

$$\longrightarrow \bigvee w^{000} \bigwedge x^0 A(x, wx)$$

But (1) contradicts the result of Feferman and Levy [7] that ($(AC)^{00}$ is, but) provided ZF is consistent then $(AC)^{01}$ is not derivable in ZF.

Remark:

Concerning $\alpha \equiv 0$ so by Howard-Kreisel [19], p. 348, 353 $(BI)_D^0$, $(AC)^{00}$ are classically deductive equivalent and near to $(C)^0$.

As described in the introduction at present the following two foundations for classical analysis exist.

Firstly, by (3.14) classical analysis is proof-theoretically equivalent to Heyting-arithmetic plus $(C)^0$. Meanwhile this has been improved in that all "local" consistency proofs of the former can be given by a uniform method in the latter.

On the other hand so far accepted and formalized intuitionistic analysis (with $(BI)_D^0$) is weaker than Π_1^1-analysis (Kreisel [27], 151). After Howard [18] (see (VII)) only Heyting-analysis plus (MP), $(\xrightarrow{\bigvee})$ and $(BI)_M^{\alpha}$

for all types α (where the higher types however are motivated more combinatorial than intuitive) is of the same proof-theoretic strength as classical analysis. By means of functional interpretation we reduced this here to Heyting-arithmetic plus T and $(hsBI)_D^{oo}$, i.e. to principles which have a constructive intuitive meaning. Thus Spector's intention to justify the passing to higher types inductively is settled.

By the preceding however comprehension and generalized bar induction are incomparable over functional interpretable domains which can be expressed in ZF. In our opinion the here given reduction to inductive processes which attaches itself to present proof theory and intuitionism is more satisfactory than the recourse to full impredicative comprehension.

XV. References

[1] Brouwer, L.E.J.: Zur Begründung der intuitionistischen Mathematik III. Math. Annalen 96 (1927), 451-488

[2] -: Zum freien Werden von Mengen und Funktionen. Indag. Math. 4 (1942), 107-108

[3] -: Points and spaces. Canadian J. Math. 6 (1954), 1-17

[4] Diller, J.: Zur Berechenbarkeit primitiv-rekursiver Funktionale endlicher Typen. Contributions to mathematical logic, Proc. Hannover Colloqu. 1966, Amsterdam 1968, 109-120

[5] -: Zur Theorie rekursiver Funktionale höherer Typen. Habilitationsschrift Munich 1968

[6] Diller, J. and Nahm, W.: Eine Variante zur Gödel-Interpretation der Heyting-Arithmetik endlicher Typen. 1971, mimeographed.

[7] Feferman, S. and Levy, A.: Independence results in set theory by Cohen's method II. Notices AMS 10 (1963), 593

[8] Gandy, R.O.: On the axiom of extensionality I, II. J. symb. logic 21 (1956), 36-48; 24 (1959), 287-300

[9] Gentzen, G.: Die Widerspruchsfreiheit der reinen Zahlentheorie. Math. Annalen 112 (1936), 493-565

[10] Girard, J.-Y.: Une extension de l'interprétation de Gödel à l'analyse, et son application à l'élimination des coupures dans l'analyse et la théorie des types. Proc. 2nd Scandinavian Logic Symposium, Amsterdam 1971, 63-92

[11] Gödel, K.: Zur intuitionistischen Arithmetik und Zahlentheorie. Ergebnisse math. Kolloqu. Heft 4 (1931-32), 34-38

[12] -: Über eine bisher noch nicht benützte Erweiterung des finiten Standpunktes, Dialectica 12 (1958), 280-287

[13] Hessenberg, G.: Kettentheorie und Wohlordnung. J. f. reine u. angew. Math. 135 (1909), 81-133, 318

[14] Heyting, A.: Infinitistic methods from a finitist point of view. Infinitistic methods, Warszawa 1959, 185-192

[15] Hilbert, D.: Über das Unendliche. Math. Annalen 95 (1926), 161-190

[16] Hilbert, D. and Bernays, P.: Grundlagen der Mathematik II. Springer 1939

[17] Hinata, S.: Calculability of primitive recursive functionals of finite type. Science reports Tokyo Kyoiki Daigaku 9 (1968), 218-235

[18] Howard, W.A.: Functional interpretation of bar induction by bar recursion. Compositio Mathematica 20 (1968), 107-124

[19] Howard, W.A. and Kreisel, G.: Transfinite induction and bar induction of types zero and one and the role of continuity in intuitionistic analysis.
J. symb. logic 31 (1966), 325-358

[20] Kleene, S.C.: Countable functionals. Constructivity in mathematics, Proc. Colloqu. Amsterdam 1957, Amsterdam 1959, 81-100

[21] -: Introduction to metamathematics. Amsterdam 1962

[22] Kleene, S.C. and Vesley, R.E.: The foundations of intuitionistic mathematics. Amsterdam 1965

[23] Kreisel, G.: Interpretation of analysis by means of constructive functionals of finite types. Constructivity in mathematics, Proc. Colloqu. Amsterdam 1957, Amsterdam 1959, 101-128

[24] -: Elementary completeness properties of intuitionistic logic with a note on negations of prenex formulae.
J. symb. logic 23 (1958), 317-330

[25] -: The non-derivability of $\neg(x)A(x)\rightarrow(Ex)\neg A(x)$, A primitive recursive, in intuitionistic formal systems.
J. symb. logic 23 (1958), 456-457

[26] -: Inessential extensions of Heyting's arithmetic by means of functionals of finite type.
J. symb. logic 24 (1959), 284

[27] -: Functions, ordinals, species. Proc. 3rd Int. Congr. Amsterdam 1967, Amsterdam 1968, 145-159

[28] Kreisel, G.: A survey of proof theory. J. symb. logic 33 (1968), 321-388

[29] -: A survey of proof theory II. Proc. 2nd Scandinavian Logic Symposium, Amsterdam 1971, 109-170

[30] Kuratowski, C.: Une méthode d'élimination des nombres transfinis des raisonnements mathématiques. Fund. Math. 3 (1922), 76-108

[31] Kuroda, S.: Intuitionistische Untersuchungen der formalistischen Logik. Nagoya Math. J. 2 (1951), 35-47

[32] Martin-Löf, P.: Hauptsatz for the theory of species. Proc. 2nd Scandinavian Logic Symposium, Amsterdam 1971, 217-233

[33] Osswald, H.: Vollständigkeit und Schnittelimination in der intuitionistischen Typenlogik. manuscripta mathematica 6 (1972), 17-31

[34] Prawitz, D.: Completeness and Hauptsatz for second order logic. Theoria 33 (1967), 246-258

[35] -: Hauptsatz for higher order logic. J. symb. logic 33 (1968), 452-457.

[36] -: Some results for intuitionistic logic with second order quantification rules. Intuitionism and proof theory, Amsterdam 1970, 259-270

[37] -: Ideas and results in proof theory. Proc. 2nd Scandinavian Logic Symposium, Amsterdam 1971, 235-307

[38] Scarpellini, B.: A model for bar recursion of higher types. Notices AMS 17 (1970), 675; Compositio Mathematica 23 (1971), 123-153

[39] -: Proof theory and intuitionistic systems. Lecture Note in Mathematics 212 (1971), Springer

[40] -: A formally constructive model for bar recursion of higher types. To appear.

[41] Schmidt, H. Arnold: Mathematische Grundlagenforschung. Enzyklop. d. math. Wissensch., neue Aufl., I/1, Heft 1, Teil II, 1948

[42] Schmidt, H. Arnold: Mathematische Gesetze der Logik I. Springer 1960

[43] Schönfinkel, M.: Über die Bausteine der mathematischen Logik. Math. Annalen 92 (1924), 305-316

[44] Schütte, K.: Die Elimination des bestimmten Artikels in Kodifikaten der Analysis. Math. Annalen 123 (1951), 166-186

[45] -: Syntactical and semantical properties of simple type theory. J. symb. logic 25 (1960), 305-326

[46] -: Beweistheorie. Springer 1960

[47] -: Grundlagen der Analysis im Rahmen einer einfachen Typenlogik. Munich 1966/67

[48] -: Theorie der Funktionale endlicher Typen. Munich 1969

[49] Shoenfield, J.R.: Mathematical logic. Addison-Wesley 1967

[50] Spector, C.: Provably recursive functionals of analysis: a consistency proof of analysis by an extension of principles formulated in current intuitionistic mathematics. Proc. Symp. Pure Math. AMS V (1962), 1-27

[51] Stanford report. Seminar on foundations of analysis. 1963, mimeographed.

[52] Tait, W.W.: Infinitely long terms of transfinite type. Formal systems and recursive functions, Amsterdam 1965, 176-185

[53] -: A nonconstructive proof of Gentzen's Hauptsatz for second order predicate logic. Bulletin AMS 72 (1966), 980-983

[54] -: Intensional interpretations of functionals of finite type I. J. symb. logic 32 (1967), 198-212

[55] -: Constructive reasoning. Proc. 3rd Int. Congr. Amsterdam 1967, Amsterdam 1968, 185-199

[56] -: Normal form theorem for bar recursive functions of finite type. Proc. 2nd Scandinavian Logic Symposium, Amsterdam 1971, 353-367

[57] Takahashi, M.: A proof of cut-elimination in simple type theory. J. Math. Soc. Japan 19 (1967), 399-410

[58] Takahashi, M.: A system of simple type theory of Gentzen-style with inference on extensionality, and the cut-elimination in it.
Commentarii Math. Univ. St. Pauli 18 (1970), 129-147

[59] -: Cut-elimination theorem and Brouwerian-valued models for intuitionistic type theory. Ibid. 19 (1971), 55-72

[60] Takeuti, G.: On a generalized logic calculus.
Japanese J. Math. 23 (1953), 39-96; 24 (1954), 149-156

[61] -: Consistency proof of subsystems of classical analysis.
Annals Math. 86 (1967), 299-348

[62] Tarski, A.: A lattice-theoretical fixpoint theorem and its applications. Pacific J. Math. 5 (1955), 285-309

[63] Troelstra, A.S.: Principles of intuitionism.
Lecture Notes in Math. 95 (1969), Springer

[64] Yasugi, M.: Intuitionistic analysis and Gödel's interpretation.
J. Math. Soc. Japan 15 (1963), 101-112

Lecture Notes in Mathematics

Comprehensive leaflet on request

Vol. 111: K. H. Mayer, Relationen zwischen charakteristischen Zahlen. III, 99 Seiten. 1969. DM 16,-

Vol. 112: Colloquium on Methods of Optimization. Edited by N. N. Moiseev. IV, 293 pages. 1970. DM 18,-

Vol. 113: R. Wille, Kongruenzklassengeometrien. III, 99 Seiten. 1970. DM 16,-

Vol. 114: H. Jacquet and R. P. Langlands, Automorphic Forms on GL (2). VII, 548 pages. 1970. DM 24,-

Vol. 115: K. H. Roggenkamp and V. Huber-Dyson, Lattices over Orders I. XIX, 290 pages. 1970. DM 18,-

Vol. 116: Séminaire Pierre Lelong (Analyse) Année 1969. IV, 195 pages. 1970. DM 16,-

Vol. 117: Y. Meyer, Nombres de Pisot, Nombres de Salem et Analyse Harmonique. 63 pages. 1970. DM 16,-

Vol. 118: Proceedings of the 15th Scandinavian Congress, Oslo 1968. Edited by K. E. Aubert and W. Ljunggren. IV, 162 pages. 1970. DM 16,-

Vol. 119: M. Raynaud, Faisceaux amples sur les schémas en groupes et les espaces homogènes. III, 219 pages. 1970. DM 16,-

Vol. 120: D. Siefkes, Büchi's Monadic Second Order Successor Arithmetic. XII, 130 Seiten. 1970. DM 16,-

Vol. 121: H. S. Bear, Lectures on Gleason Parts. III, 47 pages. 1970. DM 16,-

Vol. 122: H. Zieschang, E. Vogt und H.-D. Coldewey, Flächen und ebene diskontinuierliche Gruppen. VIII, 203 Seiten. 1970. DM 16,-

Vol. 123: A. V. Jategaonkar, Left Principal Ideal Rings. VI, 145 pages. 1970. DM 16,-

Vol. 124: Séminare de Probabilités IV. Edited by P. A. Meyer. IV, 282 pages. 1970. DM 20,-

Vol. 125: Symposium on Automatic Demonstration. V, 310 pages. 1970. DM 20,-

Vol. 126: P. Schapira, Théorie des Hyperfonctions. XI, 157 pages. 1970. DM 16,-

Vol. 127: I. Stewart, Lie Algebras. IV. 97 pages. 1970. DM 16,-

Vol. 128: M. Takesaki, Tomita's Theory of Modular Hilbert Algebras and its Applications. II, 123 pages. 1970. DM 16,-

Vol. 129: K. H. Hofmann, The Duality of Compact Semigroups and C*- Bigebras. XII, 142 pages. 1970. DM 16,-

Vol. 130: F. Lorenz, Quadratische Formen über Körpern. II, 77 Seiten. 1970. DM 16,-

Vol. 131: A Borel et al., Seminar on Algebraic Groups and Related Finite Groups. VII, 321 pages. 1970. DM 22,-

Vol. 132: Symposium on Optimization. III, 348 pages. 1970. DM 22,-

Vol. 133: F. Topsøe, Topology and Measure. XIV, 79 pages. 1970. DM 16,-

Vol. 134: L. Smith, Lectures on the Eilenberg-Moore Spectral Sequence. VII, 142 pages. 1970. DM 16,-

Vol. 135: W. Stoll, Value Distribution of Holomorphic Maps into Compact Complex Manifolds. II, 267 pages. 1970. DM 18,-

Vol. 136: M. Karoubi et al., Séminaire Heidelberg-Saarbrücken-Strasbourg sur la K-Théorie. IV, 264 pages. 1970. DM 18,-

Vol. 137: Reports of the Midwest Category Seminar IV. Edited by S. MacLane. III, 139 pages. 1970. DM 16,-

Vol. 138: D. Foata et M. Schützenberger, Théorie Géométrique des Polynômes Eulériens. V, 94 pages. 1970. DM 16,-

Vol. 139: A. Badrikian, Séminaire sur les Fonctions Aléatoires Linéaires et les Mesures Cylindriques. VII, 221 pages. 1970. DM 18,-

Vol. 140: Lectures in Modern Analysis and Applications II. Edited by C. T. Taam. VI, 119 pages. 1970. DM 16,-

Vol. 141: G. Jameson, Ordered Linear Spaces. XV, 194 pages. 1970. DM 16,-

Vol. 142: K. W. Roggenkamp, Lattices over Orders II. V, 388 pages. 1970. DM 22,-

Vol. 143: K. W. Gruenberg, Cohomological Topics in Group Theory. XIV, 275 pages. 1970. DM 20,-

Vol. 144: Seminar on Differential Equations and Dynamical Systems, II. Edited by J. A. Yorke. VIII, 268 pages. 1970. DM 20,-

Vol. 145: E. J. Dubuc, Kan Extensions in Enriched Category Theory. XVI, 173 pages. 1970. DM 16,-

Vol. 146: A. B. Altman and S. Kleiman, Introduction to Grothendieck Duality Theory. II, 192 pages. 1970. DM 18,-

Vol. 147: D. E. Dobbs, Cech Cohomological Dimensions for Commutative Rings. VI, 176 pages. 1970. DM 16,-

Vol. 148: R. Azencott, Espaces de Poisson des Groupes Localement Compacts. IX, 141 pages. 1970. DM 16,-

Vol. 149: R. G. Swan and E. G. Evans, K-Theory of Finite Groups and Orders. IV, 237 pages. 1970. DM 20,-

Vol. 150: Heyer, Dualität lokalkompakter Gruppen. XIII, 372 Seiten. 1970. DM 20,-

Vol. 151: M. Demazure et A. Grothendieck, Schémas en Groupes I. (SGA 3). XV, 562 pages. 1970. DM 24,-

Vol. 152: M. Demazure et A. Grothendieck, Schémas en Groupes II. (SGA 3). IX, 654 pages. 1970. DM 24,-

Vol. 153: M. Demazure et A. Grothendieck, Schémas en Groupes III. (SGA 3). VIII, 529 pages. 1970. DM 24,-

Vol. 154: A. Lascoux et M. Berger, Variétés Kähleriennes Compactes. VII, 83 pages. 1970. DM 16,-

Vol. 155: Several Complex Variables I, Maryland 1970. Edited by J. Horváth. IV, 214 pages. 1970. DM 18,-

Vol. 156: R. Hartshorne, Ample Subvarieties of Algebraic Varieties. XIV, 256 pages. 1970. DM 20,-

Vol. 157: T. tom Dieck, K. H. Kamps und D. Puppe, Homotopietheorie. VI, 265 Seiten. 1970. DM 20,-

Vol. 158: T. G. Ostrom, Finite Translation Planes. IV. 112 pages. 1970. DM 16,-

Vol. 159: R. Ansorge und R. Hass. Konvergenz von Differenzenverfahren für lineare und nichtlineare Anfangswertaufgaben. VIII, 145 Seiten. 1970. DM 16,-

Vol. 160: L. Sucheston, Constributions to Ergodic Theory and Probability. VII, 277 pages. 1970. DM 20,-

Vol. 161: J. Stasheff, H-Spaces from a Homotopy Point of View. VI, 95 pages. 1970. DM 16,-

Vol. 162: Harish-Chandra and van Dijk, Harmonic Analysis on Reductive p-adic Groups. IV, 125 pages. 1970. DM 16,-

Vol. 163: P. Deligne, Equations Differentielles à Points Singuliers Reguliers. III, 133 pages. 1970. DM 16,-

Vol. 164: J. P. Ferrier, Seminaire sur les Algebres Complètes. II, 69 pages. 1970. DM 16,-

Vol. 165: J. M. Cohen, Stable Homotopy. V, 194 pages. 1970. DM 16,-

Vol. 166: A. J. Silberger, PGL_2 over the p-adics: its Representations, Spherical Functions, and Fourier Analysis. VII, 202 pages. 1970. DM 18,-

Vol. 167: Lavrentiev, Romanov and Vasiliev, Multidimensional Inverse Problems for Differential Equations. V, 59 pages. 1970. DM 16,-

Vol. 168: F. P. Peterson, The Steenrod Algebra and its Applications: A conference to Celebrate N. E. Steenrod's Sixtieth Birthday. VII, 317 pages. 1970. DM 22,-

Vol. 169: M. Raynaud, Anneaux Locaux Henséliens. V, 129 pages. 1970. DM 16,-

Vol. 170: Lectures in Modern Analysis and Applications III. Edited by C. T. Taam. VI, 213 pages. 1970. DM 18,-

Vol. 171: Set-Valued Mappings, Selections and Topological Properties of 2^X. Edited by W. M. Fleischman. X, 110 pages. 1970. DM 16,-

Vol. 172: Y.-T. Siu and G. Trautmann, Gap-Sheaves and Extension of Coherent Analytic Subsheaves. V, 172 pages. 1971. DM 16,-

Vol. 173: J. N. Mordeson and B. Vinograde, Structure of Arbitrary Purely Inseparable Extension Fields. IV, 138 pages. 1970. DM 16,-

Vol. 174: B. Iversen, Linear Determinants with Applications to the Picard Scheme of a Family of Algebraic Curves. VI, 69 pages. 1970. DM 16,-

Vol. 175: M. Brelot, On Topologies and Boundaries in Potential Theory. VI, 176 pages. 1971. DM 18,-

Vol. 176: H. Popp, Fundamentalgruppen algebraischer Mannigfaltigkeiten. IV, 154 Seiten. 1970. DM 16,-

Vol. 177: J. Lambek, Torsion Theories, Additive Semantics and Rings of Quotients. VI, 94 pages. 1971. DM 16,-

Please turn over

Vol. 178: Th. Bröcker und T. tom Dieck, Kobordismentheorie. XVI, 191 Seiten. 1970. DM 18,-

Vol. 179: Seminaire Bourbaki - vol. 1968/69. Exposés 347-363. IV. 295 pages. 1971. DM 22,-

Vol. 180: Séminaire Bourbaki - vol. 1969/70. Exposés 364-381. IV, 310 pages. 1971. DM 22,-

Vol. 181: F. DeMeyer and E. Ingraham, Separable Algebras over Commutative Rings. V, 157 pages. 1971. DM 16,-

Vol. 182: L. D. Baumert. Cyclic Difference Sets. VI, 166 pages. 1971. DM 16,-

Vol. 183: Analytic Theory of Differential Equations. Edited by P. F. Hsieh and A. W. J. Stoddart. VI, 225 pages. 1971. DM 20,-

Vol. 184: Symposium on Several Complex Variables, Park City, Utah, 1970. Edited by R. M. Brooks. V, 234 pages. 1971. DM 20,-

Vol. 185: Several Complex Variables II, Maryland 1970. Edited by J. Horváth. III, 287 pages. 1971. DM 24,-

Vol. 186: Recent Trends in Graph Theory. Edited by M. Capobianco/ J. B. Frechen/M. Krolik. VI, 219 pages. 1971. DM 18,-

Vol. 187: H. S. Shapiro, Topics in Approximation Theory. VIII, 275 pages. 1971. DM 22,-

Vol. 188: Symposium on Semantics of Algorithmic Languages. Edited by E. Engeler. VI, 372 pages. 1971. DM 26,-

Vol. 189: A. Weil, Dirichlet Series and Automorphic Forms. V. 164 pages. 1971. DM 16,-

Vol. 190: Martingales. A Report on a Meeting at Oberwolfach, May 17-23, 1970. Edited by H. Dinges. V, 75 pages. 1971. DM 16,-

Vol. 191: Séminaire de Probabilités V. Edited by P. A. Meyer. IV, 372 pages. 1971. DM 26,-

Vol. 192: Proceedings of Liverpool Singularities - Symposium I. Edited by C. T. C. Wall. V, 319 pages. 1971. DM 24,-

Vol. 193: Symposium on the Theory of Numerical Analysis. Edited by J. Ll. Morris. VI, 152 pages. 1971. DM 16,-

Vol. 194: M. Berger, P. Gauduchon et E. Mazet. Le Spectre d'une Variété Riemannienne. VII, 251 pages. 1971. DM 22,-

Vol. 195: Reports of the Midwest Category Seminar V. Edited by J.W. Gray and S. Mac Lane.III, 255 pages. 1971. DM 22,-

Vol. 196: H-spaces - Neuchâtel (Suisse)- Août 1970. Edited by F. Sigrist, V, 156 pages. 1971. DM 16,-

Vol. 197: Manifolds - Amsterdam 1970. Edited by N. H. Kuiper. V, 231 pages. 1971. DM 20,-

Vol. 198: M. Hervé, Analytic and Plurisubharmonic Functions in Finite and Infinite Dimensional Spaces. VI, 90 pages. 1971. DM 16,-

Vol. 199: Ch. J. Mozzochi, On the Pointwise Convergence of Fourier Series. VII, 87 pages. 1971. DM 16,-

Vol. 200: U. Neri, Singular Integrals. VII, 272 pages. 1971. DM 22,-

Vol. 201: J. H. van Lint, Coding Theory. VII, 136 pages. 1971. DM 16,-

Vol. 202: J. Benedetto, Harmonic Analysis on Totally Disconnected Sets. VIII, 261 pages. 1971. DM 22,-

Vol. 203: D. Knutson, Algebraic Spaces. VI, 261 pages. 1971. DM 22,-

Vol. 204: A. Zygmund, Intégrales Singulières. IV, 53 pages. 1971. DM 16,-

Vol. 205: Séminaire Pierre Lelong (Analyse) Année 1970. VI, 243 pages. 1971. DM 20,-

Vol. 206: Symposium on Differential Equations and Dynamical Systems. Edited by D. Chillingworth. XI, 173 pages. 1971. DM 16,-

Vol. 207: L. Bernstein, The Jacobi-Perron Algorithm - Its Theory and Application. IV, 161 pages. 1971. DM 16,-

Vol. 208: A. Grothendieck and J. P. Murre, The Tame Fundamental Group of a Formal Neighbourhood of a Divisor with Normal Crossings on a Scheme. VIII, 133 pages. 1971. DM 16,-

Vol. 209: Proceedings of Liverpool Singularities Symposium II. Edited by C. T. C. Wall. V, 280 pages. 1971. DM 22,-

Vol. 210: M. Eichler, Projective Varieties and Modular Forms. III, 118 pages. 1971. DM 16,-

Vol. 211: Théorie des Matroïdes. Edité par C. P. Bruter. III, 108 pages. 1971. DM 16,-

Vol. 212: B. Scarpellini, Proof Theory and Intuitionistic Systems. VII, 291 pages. 1971. DM 24,-

Vol. 213: H. Hogbe-Nlend, Théorie des Bornologies et Applications. V, 168 pages. 1971. DM 18,-

Vol. 214: M. Smorodinsky, Ergodic Theory, Entropy. V, 64 pages. 1971. DM 16,-

Vol. 215: P. Antonelli, D. Burghelea and P. J. Kahn, The Concordance-Homotopy Groups of Geometric Automorphism Groups. X, 140 pages. 1971. DM 16,-

Vol. 216: H. Maaß, Siegel's Modular Forms and Dirichlet Series. VII, 328 pages. 1971. DM 20,-

Vol. 217: T. J. Jech, Lectures in Set Theory with Particular Emphasis on the Method of Forcing. V, 137 pages. 1971. DM 16,-

Vol. 218: C. P. Schnorr, Zufälligkeit und Wahrscheinlichkeit. IV, 212 Seiten 1971. DM 20,-

Vol. 219: N. L. Alling and N. Greenleaf, Foundations of the Theory of Klein Surfaces. IX, 117 pages. 1971. DM 16,-

Vol. 220: W. A. Coppel, Disconjugacy. V, 148 pages. 1971. DM 16,-

Vol. 221: P. Gabriel und F. Ulmer, Lokal präsentierbare Kategorien. V, 200 Seiten. 1971. DM 18,-

Vol. 222: C. Meghea, Compactification des Espaces Harmoniques. III, 108 pages. 1971. DM 16,-

Vol. 223: U. Felgner, Models of ZF-Set Theory. VI, 173 pages. 1971. DM 16,-

Vol. 224: Revêtements Etales et Groupe Fondamental. (SGA 1). Dirigé par A. Grothendieck XXII, 447 pages. 1971. DM 30,-

Vol. 225: Théorie des Intersections et Théorème de Riemann-Roch. (SGA 6). Dirigé par P. Berthelot, A. Grothendieck et L. Illusie. XII, 700 pages. 1971. DM 40,-

Vol. 226: Seminar on Potential Theory, II. Edited by H. Bauer. IV, 170 pages. 1971. DM 18,-

Vol. 227: H. L. Montgomery, Topics in Multiplicative Number Theory. IX, 178 pages. 1971. DM 18,-

Vol. 228: Conference on Applications of Numerical Analysis. Edited by J. Ll. Morris. X, 358 pages. 1971. DM 26,-

Vol. 229: J. Väisälä, Lectures on n-Dimensional Quasiconformal Mappings. XIV, 144 pages. 1971. DM 16,-

Vol. 230: L. Waelbroeck, Topological Vector Spaces and Algebras. VII, 158 pages. 1971. DM 16,-

Vol. 231: H. Reiter, L^1-Algebras and Segal Algebras. XI, 113 pages. 1971. DM 16,-

Vol. 232: T. H. Ganelius, Tauberian Remainder Theorems. VI, 75 pages. 1971. DM 16,-

Vol. 233: C. P. Tsokos and W. J. Padgett. Random Integral Equations with Applications to Stochastic Systems. VII, 174 pages. 1971. DM 18,-

Vol. 234: A. Andreotti and W. Stoll. Analytic and Algebraic Dependence of Meromorphic Functions. III, 390 pages. 1971. DM 26,-

Vol. 235: Global Differentiable Dynamics. Edited by O. Hájek, A. J. Lohwater, and R. McCann. X, 140 pages. 1971. DM 16,-

Vol. 236: M. Barr, P. A. Grillet, and D. H. van Osdol. Exact Categories and Categories of Sheaves. VII, 239 pages. 1971. DM 20,-

Vol. 237: B. Stenström. Rings and Modules of Quotients. VII, 136 pages. 1971. DM 16,-

Vol. 238: Der kanonische Modul eines Cohen-Macaulay-Rings. Herausgegeben von Jürgen Herzog und Ernst Kunz. VI, 103 Seiten. 1971. DM 16,-

Vol. 239: L. Illusie, Complexe Cotangent et Déformations I. XV, 355 pages. 1971. DM 26,-

Vol. 240: A. Kerber, Representations of Permutation Groups I. VII, 192 pages. 1971. DM 18,-

Vol. 241: S. Kaneyuki, Homogeneous Bounded Domains and Siegel Domains. V, 89 pages. 1971. DM 16,-

Vol. 242: R. R. Coifman et G. Weiss, Analyse Harmonique Non-Commutative sur Certains Espaces. V, 160 pages. 1971. DM 16,-

Vol. 243: Japan-United States Seminar on Ordinary Differential and Functional Equations. Edited by M. Urabe. VIII, 332 pages. 1971. DM 26,-

Vol. 244: Séminaire Bourbaki - vol. 1970/71. Exposés 382-399. IV, 356 pages. 1971. DM 26,-

Vol. 245: D. E. Cohen, Groups of Cohomological Dimension One. V, 99 pages. 1972. DM 16,-